ATM Networks

ATM Networks

Principles and Use

Martin P. Clark

Netro Corporation
Santa Clara, California, USA
and
Frankfurt, Germany

WILEY TEUBNER

A Partnership between John Wiley & Sons and B. G. Teubner Publishers

Chichester · New York · Brisbane · Toronto · Singapore · Stuttgart · Leipzig

National 01243 779777
International (+44) 1243 779777

Reprinted March 1997

Other Wiley Editorial Offices

John Wiley & Sons, Inc., 605 Third Avenue,
New York, NY 10158-0012, USA

Jacaranda Wiley Ltd, 33 Park Road, Milton,
Queensland 4064, Australia

John Wiley & Sons (Canada) Ltd, 22 Worcester Road,
Rexdale, Ontario M9W 1L1, Canada

John Wiley & Sons (Asia) Pte Ltd, 2 Clementi Loop #02-01,
Jin Xing Distripark, Singapore 0512

Die Deutsche Bibliothek – CIP Einheitsaufnahme

Clark, Martin P.:
ATM networks: principles and use/Martin P. Clark, –
Chichester; New York; Brisbane; Toronto; Singapore;
Stuttgart; Leipzig: Wiley-Teubner, 1996
ISBN 3 519 06448 0 (Teubner)
ISBN 0 471 96701 7 (Wiley)

Library of Congress Cataloguing-in-Publication Data

Clark, Martin P.
ATM networks: principles and use / Martin P. Clark.
p. cm.
Includes bibliographical references and index.
ISBN 0-471-96701-7
1. Asynchronous transfer mode. 2. Integrated services digital networks. 3. Broadband communication systems. I. Title.
TK5105.35.C52 1996
004.6'6—dc20 96-11992
CIP

British Library Cataloguing in Publication Data

A catalogue record for this book is available from the British Library

ISBN 0 471 96701 7

Typeset in 10/12pt Times by Vision Typesetting, Manchester
Printed and bound in Great Britain by Biddles Ltd, Guildford and King's Lynn
This book is printed on acid-free paper responsibly manufactured from sustainable forestation, for which at least two trees are planted for each one used for paper production.

Contents

Preface

The modern telecommunications world is a complicated place, overloaded with jargon and bewilderingly full of different technical options and opportunities. Many experts are only experts of small domains and within limited geographies. For newcomers it is often difficult to gain a grasp on the basic principles and even harder to unravel the mysterious technical paradoxes and the apparent contradictions of many modern technologies.

My greatest hope in writing this book is that you, the reader, will find help and insight amongst its pages. I wish for no greater commendation than your thoughts that you find the subject of ATM presented here in an accessible and readable form. I hope also that the comprehensive glossary will remain with you, as your wayfinder through all that jargon!

Martin P. Clark
Frankfurt, Germany
20 December 1995

Acknowledgements

No book on Asynchronous Transfer Mode (ATM) could fail to recognize the invaluable contribution to this technology and to world standardization as a whole made by the International Telecommunications Union and the ATM Forum, and you will find references to their work throughout the text. Particular copyright extracts are labelled accordingly, but the full texts may be obtained (as relevant) from ITU Sales and Marketing Service, Place des Nations, CH-1211 Geneva 20, Switzerland or from ATM Forum, 2570 West El Camino Real, Suite 304, Mountain View, California CA 94040, USA. Alternatively, they may be contacted over Internet respectively at sales@itu.ch and info@atmforum.com.

1

Introduction to ATM

ATM, or Asynchronous Transfer Mode, is the most modern of telecommunications switching techniques, and is at the core of the future Broadband Integrated Services Digital Network (B-ISDN). In this chapter we introduce them both.

1.1 What are ATM and B-ISDN?

What does ATM stand for? It stands for *Asynchronous Transfer Mode.* And what is it? It is the most modern of telecommunications switching techniques, a highly efficient switching technique which is able to switch connections for a wide range of different information types at a wide range of different bitrates (capacities). It is not a switching technique limited to telephone switching or to data networking, but rather a technique which allows a network to be used simultaneously for the transfer of different signal types (e.g. telephone, data, video, etc.). It is an *integrated* switching technique. It is *the* integrated switching technique which will form the basis of the *Broadband Integrated Services Digital Network (B-ISDN).*

So what is the B-ISDN? The B-ISDN is the most modern type of telecommunications network – one offering simultaneous switching of different information types, for the carriage of *multimedia* applications. But not just that. The B-ISDN is the most intelligent of *intelligent networks (INs)*. It has built into it powerful network control systems, capable of supporting sophisticated network services and network management operations. Thus, for example, connections could be set up not only according to the dialled number, but also by evaluating the identity of the caller, his creditworthiness, the time of day and the current network loading. For some types of connections, the call charges may be accrued to the called party (*freephone* or *800* service), or charged at premium rate according to the content value of the information (e.g. videofilm or database information) received by the caller. In short, B-ISDN is the network for everything!

What then, is the difference between ATM and B-ISDN? The terms may often appear to be used synonymously, and correctly so, for ATM is a subset of B-ISDN. ATM is the *switching* technique at the heart of B-ISDN. B-ISDN is not only ATM, but a complete network and management control architecture. Thus when speaking about the switching capabilities of B-ISDN, one can speak interchangeably about B-ISDN or ATM switching capabilities.

In this book the prime focus is on ATM, but we also discuss in depth the various design aspects and technical standards concerning B-ISDN. The two are simply inseparable. The focus is on the principles of the technologies, on the features of their network architecture, on their strengths and on their limitations.

ATM and B-ISDN are technologies – the internal 'innards' of the most modern of telecommunications networks. They are not, in themselves, telecommunications *services* which an end-user may recognize, but rather the means for carrying the various users' information. And in the same way that today you would probably buy a *Pentium* or *Power-PC* based personal computer because of a layman's understanding of the value of greater processing speed, so you would be well advised to base your new telecommunications network around ATM, because of 'greater efficiency and flexibility'. Nonetheless, a more informed appreciation of the full scope of the new possibilities will help to allow you make the best of ATM. This appreciation requires a greater technical understanding. The aim of this book is to explain this technology – to try to make it understandable and accessible.

1.2 The Services Offered by a B-ISDN

The standards classify the services offered by B-ISDN networks into two categories – *interactive services* and *distribution services*. As Figure 1.1 shows, these categories are then further subdivided into five further service types.

Interactive services are normally based on communications between just two parties. There are three subtypes:

1. An example of a *conversational service* is a telephone conversation or a point-to-point data connection, where the two end-points of the communication are in 'real-time' connected with one another and thus 'converse'.

2. A *message service* is a telecommunications service similar to the postal service. A message is submitted to the network (like a letter is posted).

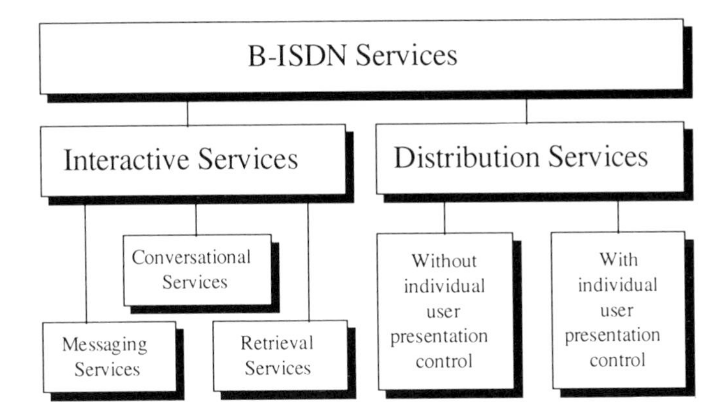

Figure 1.1 The service types offered by B-ISDN

Sometime later, the message is delivered to the given address. The recipient may read the message and reply with a *message* if he wishes, but the sender at this time is no longer connected. Message services are usually not *guaranteed*. The network is not able to check whether the recipient address is valid, and may return no confirmation to the sender of receipt. Thus 'no reply' may result either because the intended recipient never got the message or because the recipient chose not to respond.

3. A *retrieval service* is where a network caller accesses a central server, database or storage archive, requesting the delivery of certain specified information. The caller might receive data about holiday bargains (akin to *videotext services – Prestel, Minitel, Datex-J*, etc.), new video clips, etc.

Distribution services are services in which the information from a single source is distributed to many recipients at the same time. Distribution services are subdivided into those with or without individual *user presentation control*. An example of a distribution service without individual *user presentation control* is the broadcasting of national television. All the recipients receive the same signal at the same time. An example of a distribution service with individual user presentation control is *video-on-demand*. Here only those users who wish to pay and receive a given film do so.

There is intense effort in many branches of telecommunications development and research leading to many exciting potential applications of the various types to run over B-ISDN networks (Table 1.1)

Table 1.1 Potential applications of the various B-ISDN service categories

Category	Service type	Potential applications
Interactive Services	Conversational services	Voice telephony
	Messaging services	Internet Electronic Mail
	Retrieval services	On-Line Database service
Distribution Services	Without user control	Broadcast TV
	With user control	Pay-on-view TV or Video-on-demand

1.3 What Does *Asynchronous Transfer Mode* Mean?

The terminology *transfer mode* indicates that ATM is a telecommunications transport technique – a method by which information may be transferred (switched and transported) from one side of a network to the other. The term *asynchronous* distinguishes the technique from *synchronous* and *plesiochronous* transfer techniques.

Synchronous transfer mode (STM) is the method used in high speed transmission systems (such as *synchronous digital hierarchy, SDH*, or *synchronous optical network, SONET* – see Appendix 1). In a *synchronous* transfer technique, the line capacity (bitrate) is structured in a strictly regular, and repeating pattern. Thus a 155 Mbit/s SDH line transmission system, for example, is actually composed of a *frame* of 2430 bytes (correctly 2430 *octets*), repeated 8000 times per second. There are no gaps between the frames, so the same part of the frame can be expected in the same place every 125 microseconds – the system is *synchronous*. In a *plesiochronous* system, the system does not run quite synchronously, but pretends that it does. In order to circumvent the inevitable errors which occur because the system is not quite synchronous, some of the capacity is purposely wasted in order that there is slack in the system, and the end-user does not suffer the ill-effects.

In the *asynchronous transfer mode*, frames (or correctly *cells*) of information are only sent when necessary. Thus, for example, cells are only sent across the network to represent the alphabetic characters which I am typing and only when I type something. In between, nothing is sent. By comparison, synchronous transfer mode would convey frames all the time – empty ones at times.

The *asynchronous transfer mode (ATM)* is thus potentially the more efficient of the telecommunications transport techniques. We discuss the reasons further in chapter 2.

1.4 What Types of Connections Does an ATM Network or B-ISDN Support?

So what types of connection does an ATM network support? The answer, as we have seen, is all types. In Figure 1.2, we illustrate as a brief summary the main types of connections and services which ATM networks will offer. We also present some of the main interfaces of ATM (the *UNI* and the *NNI*). Lastly, the diagram shows how ATM relates to B-ISDN.

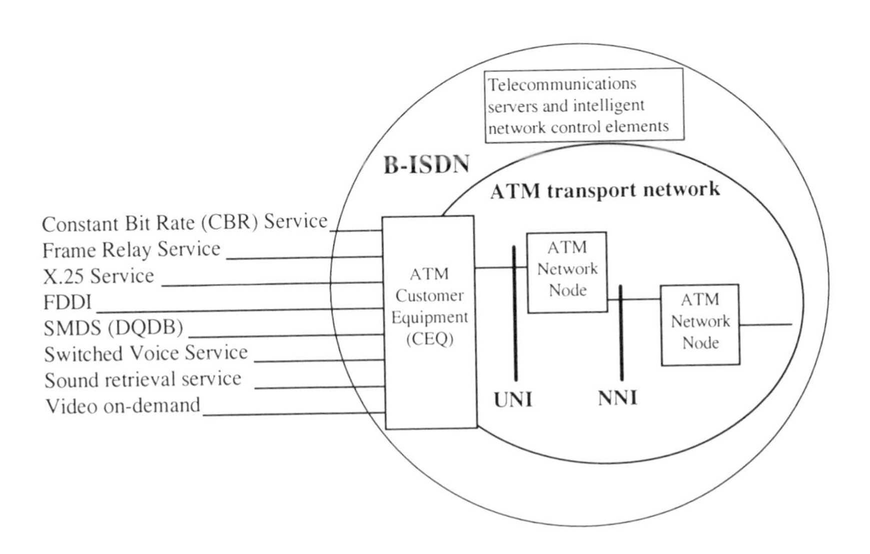

Term	Meaning
CBR	Constant Bit Rate (a point-to-point connection across an ATM network providing service similar to a *private wire* or *leaseline*)
DQDB	Dual Queue Dual Bus (The technology behind SMDS)
FDDI	Fibre Distributed Data Interface (a campus technology for interconnecting LAN routers)
NNI	Network–Node Interface (an ATM network interface)
SMDS	Switched Multimegabit Data Service (an existing broadband network type)
UNI	User–Network Interface (an ATM network interface)

Figure 1.2 ATM and B-ISDN: the connection types available

1.5 When Should I Consider Using ATM?

Before embarking today on any major new telecommunications network investments or on the development of any major new computer software applications (particularly *broadband* or *multimedia* applications), one should consider including ATM switching elements in the network architecture or ATM network interfaces in the software APIs (application programming interfaces). Why? Because ATM is likely to become *the* universal telecommunications transport technique. It will, however, take a few years yet for ATM to establish itself fully as *the* information highway, during which time the best informed corporations and network operators can steal a march on their competitors. This will require having an understanding of the technical capabilities of ATM, having faith in its potential and determining to make it happen. The remainder of this book is dedicated to helping along the way!

2

The Concept of ATM

As a foundation for understanding the reasons why ATM has developed, and as a basis for our later comparisons of the strengths of ATM compared with other telecommunications transport technologies, this chapter presents the basic technical principles and terminology, explaining what marks out ATM from its predecessors. In particular, we discuss the principles of statistical multiplexing and the specifics of cell switching.

2.1 A Flexible Transmission Medium

ATM is a telecommunications transmission technique. It is one of the most modern transmission techniques. It is designed to be the most flexible and efficient. An ATM-equipped transmission line or telecommunications network is able to support:

- usage by multiple users simultaneously, each with
- different telecommunication needs (e.g. telephone, data transmission, LAN interconnection, videotransmission, etc.) and with
- each application running at different transmission speeds (i.e. with differing bandwidth needs).

But these capabilities are also offered by predecessor technologies, so why pay the premium for an ATM network, you might ask? What distinguishes ATM from its predecessors is that it performs these functions more efficiently. In particular, ATM is capable of an instant-by-instant adjustment in the allocation of the available network capacity between the various users competing for its use. Rather than allocating fixed capacity between the two communicating parties for the duration of a call or *session*, ATM ensures that the line capacity is optimally used, by carrying only the needed, or 'useful', information. For example, the silent patches in speech need not be transmitted. In the meantime a short burst of data could be carried. The dynamic allocation of bandwidth is achieved by ATM using a newly developed technique called *cell relay switching*. Understanding its principles is the key to understanding ATM, its strengths and limitations.

2.2 *Statistical Multiplexing* and the Evolution of *Cell Relay Switching*

ATM is based upon a *statistical multiplexing* technique called *cell relay switching*. *Statistical multiplexing* is a means of multiplying the effective capacity of a transmission line or network, by taking advantage of the statistical nature of the occurrences when information needs to be carried.

When statistically multiplexing speech connections, the silent periods can be suppressed and not sent over the line. Meanwhile the words from other conversations can be carried in the gaps. The same principle can also be used for data transmission – and even more effectively. Either the separate characters of different texts can be interleaved or different data files can be sent quickly one after another.

The major benefit of statistical multiplexing is that the useful carrying capacity of the line is maximized by avoiding the unnecessary transmission of redundant information (i.e. pauses). But in addition, since the full capacity of the line (i.e. its full speed) is made available for each individual connection or carriage of data information, the transmission time (propagation time) may be reduced. (Other techniques allocate fixed sub-portions of the bandwidth to the individual connections.)

In the example of Figure 2.1, we illustrate the technique of statistical multiplexing. Three separate users (represented by sources A, B and C) are to communicate over the same transmission line, sharing the line resources by means of statistical multiplexing. The three separate source circuits are fed into a statistical multiplexor, which is connected by a single line to the demultiplexor at the receiving end. (A similar arrangement, using a second line for the receive channel, but with multiplexor on the right and demultiplexor on the left, will also be necessary but this is not shown.)

The statistical multiplexor sends whatever it receives from any of the source channels directly onto the transmission line. And why is it called *statistical multiplexing*? Simply because it relies on the statistical unlikelihood of all three channels wanting to send information simultaneously. Actually, for a short

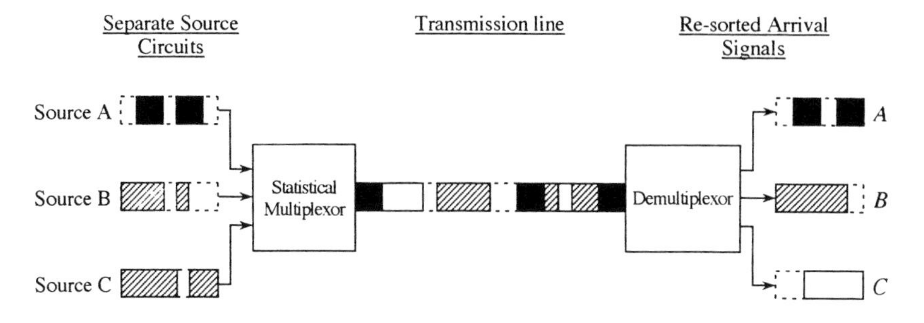

Figure 2.1 The principle of statistical multiplexing

period of time, the multiplexor is designed to be able to cope with simultaneous transmission from all of the sources. This is done by sending the most important signal directly to line and storing the least important signal in a *buffer* for an instant until the first signal has been transmitted.

Statistical multiplex systems cannot cope with prolonged periods of simultaneous transmission by two or more sources. During periods of prolonged simultaneous transmission, the storage space *(queueing buffer)* progressively fills up to capacity and starts to overflow (like a bucket overflowing).

Surprisingly, perhaps, the loss of small amounts of information may not be catastrophic. Data transmission techniques (*protocols*) are designed especially to detect the loss of information and arrange for retransmission of the information from the source. And even for signals not automatically transmitted (e.g. the human voice), the loss may only result in a slight 'click' being heard by the listener or complete unawareness of the loss. Prolonged periods of information loss (say more than 0.1 seconds, 100 ms) are, however, a major threat to acceptable connection quality.

In order to guard against the possibility of information loss, the system should be so planned that the sum of the *average* throughputs of each of the source channels is less than the maximum capacity of the transmission line (whereby the *average rate* should be measured over a relatively short time period). In other words, for statistical multiplexing to work reliably and acceptably, line throughput capacity must exceed the sum of average source throughputs (A+B+C). Actually the line throughput should be at least 1.5 to 2 times the sum of the average throughputs, otherwise periods of congestion are experienced, punctuated by sometimes long periods of slack.

The first practical realizations of statistical multiplexing were data networking protocols. In particular, statistical multiplexing forms the basis of data *packet switching*. It is the principle upon which IBM's SNA (systems network architecture) and ITU-T's X.25 recommendation are based. As such, statistical multiplexing is widely in use within computer data networks.

Today's public voice networks, in contrast to data networks, have not used statistical multiplexing. Instead, voice and telephone networks have historically been based upon *circuit switching* – the allocation of a path across the network on a fully dedicated point-to-point basis for the duration of the call. The strength of circuit switching is the guaranteed throughput and delay performance of the resulting connection. This is critical in order that acceptable voice quality can be achieved (in the subjective opinion of telephone users). A telephone call connected in a circuit-switched manner is like an empty pipe between two telephone users. Whatever one speaker says into the pipe comes out at the other end in an identical format – but only one pair of callers can use the pipe during any particular call.

Historically, telephone networks have not used statistical multiplexing techniques because of the problem of achieving acceptable speech quality. There were attempts to 'packet switch' voice across data networks, but the problem was that individual words or parts of words take different times to propagate through a packet network, so that the listener hears a rather broken form of the original signal. The effect is caused by a phenomenon called *jitter*. The more jitter (variable propagation delay) experienced by a telephone connection, the worse the perceived quality of the connection.

Jitter in data networks is relatively unimportant. So long as the average delay is not great, computer users do not notice whether some typed characters appear imperceptibly faster or slower than others. As a result, packet switched networking techniques have dominated the world of data transmission, since they are the most efficient in employing available transmission capacity. Meanwhile, voice networks have remained circuit switch-based, because of the quality problems. The consequence has been the evolution of two entirely separate networking worlds – voice and data. Transmission lines cannot easily and efficiently be shared between voice and data, and dynamic allocation of bandwidth – one instant to voice, the next moment to data – has not been possible.

A number of attempts have been made to develop technologies capable of handling equally well both voice and data over the same network. The two most noteworthy technologies in this category are *ISDN (integrated services digital network)* and, now, *ATM (asynchronous transfer mode)*.

ISDN is a technology based upon circuit switching within digital telephone networks. The digital nature of the network is exploited for the use of data transmission. And to make the *integrated* data and voice carriage possible, the signalling within the network (between the exchanges in the network and from the calling handset to the first exchange) is very advanced – far superior to the simple pulsing technique used in older *analogue* telephone networks. Unfortunately, ISDN as a data transport medium is limited in its efficiency and flexibility due to its circuit switched nature. This we discuss more fully in our comparison between ATM and ISDN in chapter 3.

While ISDN will revamp customer expectations of telephone services (with new features like caller identification before answer and *ring back when free*), ISDN in its basic form (*narrowband ISDN*) is unlikely to form the basis of advanced *integrated* voice and data networks. This is because of its limited capability for data carriage.

In contrast, because ATM (which is a form of so-called *broadband ISDN* or *B-ISDN)* has evolved from the statistical multiplexing technique inherent in packet switched data networks, it is likely to have more success as an integrated voice, data and video transport medium. The developers of ATM have simply concentrated on improving the packet switching technique to limit the jitter on speech, video and other delay sensitive applications. The resulting technique is called *cell relay switching*, or simply *cell relay.*

2.3 The Problems to be Solved by *Cell Relay*

The normal statistical multiplexing of data connections is carried out by *packet switching*. *Packets* of data are created by each of the sources, and interleaved as appropriate by the statistical multiplexor, as already illustrated in Figure 2.1. The interleaving is usually carried out on a simple *FIFO queue* basis (*first in–first out*). Packets received from the sending sources are stored at the back of the queue. Meanwhile, packets at the front of the queue are being transmitted to line.

A typical data packet contains typically between 1 and 256 characters (between 8 and 2048 bits), and the linespeed is typically 9600 bit/s. The propagation delay at a time when two sources try to send simultaneously (due to the extra waiting time) may therefore be up to 200 ms (2048/9600 s) longer than when there is no simultaneous transmission. In other words, there may be up to 200 ms of jitter. This is unacceptable for speech transmission. But before discussing how *cell relay* circumvents this problem, we should cover one other important aspect of statistical multiplexing – an aspect which constrains the maximum achievable line usage efficiency.

In order to allow the demultiplexor (see Figure 2.1) to sort out the various packets belonging to the different *logical connections*, and forward them to the correct destinations (A to *A*, B to *B*, C to *C* etc.) there needs to be a label attached to each packet to say to which *logical connection* (i.e. telephone conversation or data communications *session*) it belongs. This label is contained in the *header*, which is an addition to the front of the packet and has a function like the envelope of a letter. The header (Figure 2.2) is crucial to the technique of statistical multiplexing, but has the disadvantage that it

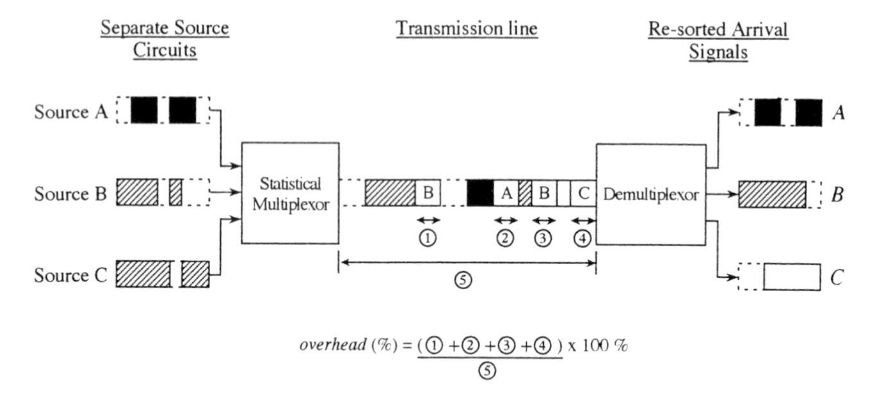

Figure 2.2 Statistical multiplexing *headers* and *overhead*

adds to the information which must be carried by the transmission line between multiplexor and demultiplexor. At the demultiplexor, the header is removed so it does not disturb the receiver, but meanwhile it has generated an *overhead* load for the transmission line. It is thus impossible using statistical multiplexing techniques to achieve 100 per cent loading of a transmission line with raw user information. Some of the capacity has to be given up to carry the *overhead*.

The major challenges for ATM developers have therefore been to minimize the jitter experienced by speech, video and other delay-sensitive applications, while simultaneously optimizing line efficiency by minimizing network *overhead*. As we shall see, these demands contend with one another.

2.4 The Technique of *Cell Relay*

Cell relay is a form of statistical multiplexing similar in many ways to *packet switching*, except that the *packets* are instead called *cells*. Each of the *cells* is of a fixed rather than a variable size.

The fixed cell size defined by ATM standards is 48 octets (bytes) plus a 5-octet header (i.e. 53 octets in all – see Figure 2.3). The transmission line speeds currently foreseen to be used are either 12, 25, 34, 45, 155, or 622 Mbit/s. We can therefore immediately draw certain conclusions about ATM performance:

- the overhead is at least 5 bytes in 53 bytes, i.e. >9 per cent;
- the duration of a cell is at most 53×8 bits/12 Mbit $s^{-1} = 36$ μs. (12 ks at 34 Mbit s^{-1}).

Since the cell duration is relatively short, provided a priority scheme is applied to allow cells from delay-sensitive signal sources (e.g. speech, video, etc.) to have access to the next cell *slot*, then the jitter (variation in signal propagation delay) can be kept very low – not zero as is possible with circuit switching, but at least low enough to give a subjectively acceptable quality for telephone listeners or video watchers. Jitter-insensitive traffic sources (e.g. datacommunication channels) can be made to wait for the allocation of the next free or *low priority slot* (Figure 2.4).

We could reduce the jitter still further by reducing the cell size, but this would increase the proportion of the line capacity needed for carrying the cell headers, and thus reduce the line efficiency.

The *cell header* carries information sufficient to allow the ATM network to determine to which connection (and thus to which destination port and end-user) each cell should be delivered. We could draw a comparison with a postal service and imagine each of the cells to be a letter with 48 characters of information contained in an envelope on which a 5-digit postcode appears. You simply drop your letters (cells) in in the right order and they come out in the same order at the other end, though maybe slightly jittered in time. Just as a postal service has numerous vans, lorries and personnel to carry different letters over different stretches, and sorting offices to direct the letters along their individual paths, so an ATM network can comprise a mesh of transmission links and switches to direct individual cells by inspecting the *address* contained in the *header* (Figure 2.5).

The ATM 'switch' acts in much the same way as a postal sorter. On its incoming side is a FIFO (first in–first out) buffer, like a pile of letters. At the

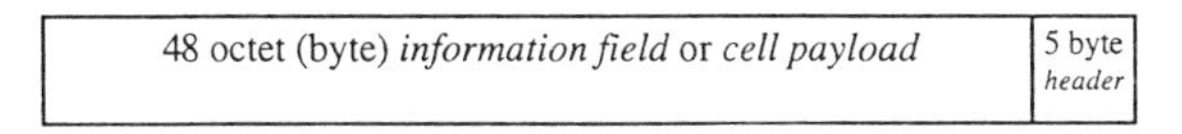

Figure 2.3 ATM 53-byte cell format

Cell	Free slot	Cell	Cell

Figure 2.4 The *cells* and *slots* of cell relay

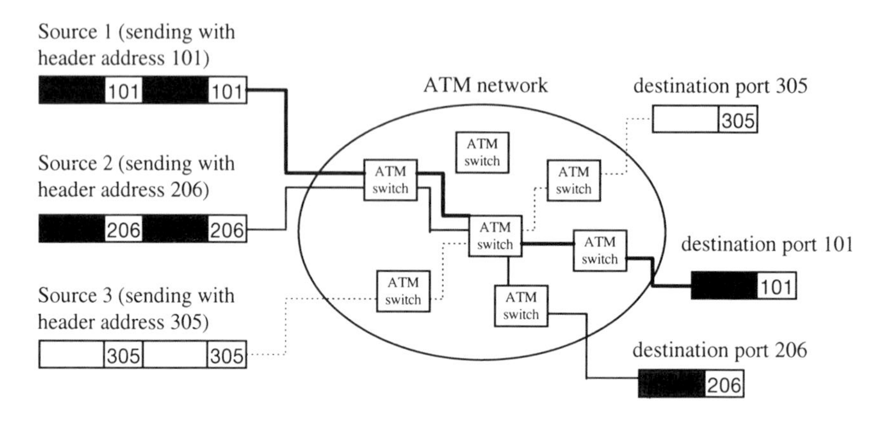

Figure 2.5 Switching in an ATM network

front of the buffer (like the top letter in the pile) is the cell which has been waiting longest to be switched. New cells arriving are added to the back of the buffer. The switching process involves looking at each cell in turn, and determining from the address held in the header which outgoing line should be taken. The cell is then added to the FIFO output buffer which is queueing cells waiting to be transmitted on this line. The cell then proceeds to the next exchange.

You may think that a 5–digit postcode is rather inadequate as a means of addressing all the likely users of an ATM network, and it might have been if not for several provisions of the ATM specifications. First, the 'digits' are whole octets (base 256) rather than decimal digits (base 10). This means that the header has the range for over 10^{12} combinations (40 bits), though only a maximum of 28 bits (2.7×10^8 combinations) are ever used for addressing. The second provision of ATM addressing is that the addresses (correctly called *identifiers*) are only allocated to active connections.

ATM connections are allocated an *identifier* during call setup, and this is reallocated to another connection when the connection is cleared. In this way the number of different identifiers available need not directly reflect the number of users connected to the network (which may be many millions), only the number of simultaneously active connections. Thirdly, as we explain in more detail later in the book, various subregions of the network may use different identifier schemes, thus multiplying the available capacity, but then demanding the ability of network nodes to *translate* (i.e. amend) identifiers in the 5-octet header.

By highly efficient usage of the information carried in the header, the overall length of the header can be kept to a minimum. As a result, the network *overhead* is minimized.

2.5 The Components of an ATM Network

There are four basic types of equipment which go to make up an ATM network. These are:

- Customer equipment (CEQ)
- ATM switches
- ATM crossconnects
- ATM multiplexors

These elements combine together to make a network as shown in Figure 2.6. A number of standard interfaces are also defined by the ATM specifications as the basis for the connections between the various components. The most important interfaces are:

- *User Network Interface (UNI)*
- *Network Node Interface (NNI)*
- *Inter-Network Interface (INI)*

These are also shown in Figure 2.6.

ATM UNI (User Network Interface)

The ATM User Network Interface (UNI) is the standard technical specification allowing ATM customer equipment (CEQ) from various different manufacturers to communicate over a network provided by yet another manufacturer. It is the interface employed between ATM customer equipment and either ATM multiplexor, ATM crossconnect or ATM switch. It consists of a set of layered protocols as we shall discuss later.

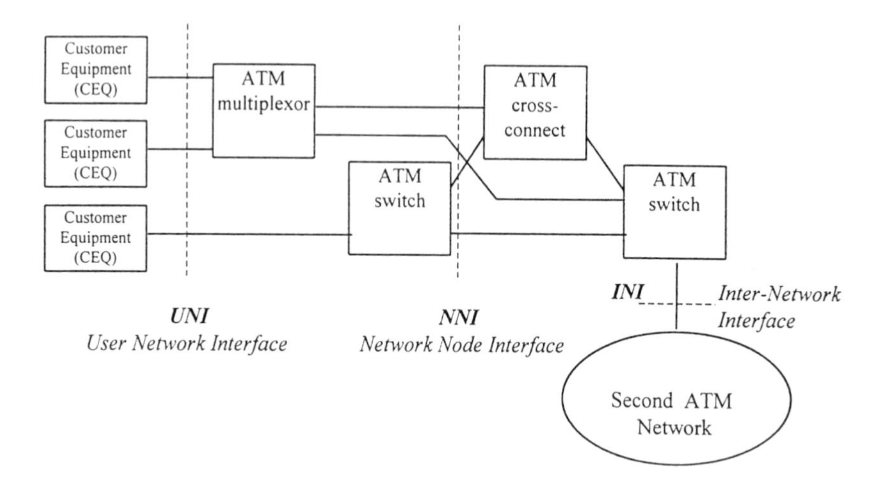

Figure 2.6 The components of an ATM network

ATM Customer Equipment (CEQ)

ATM customer equipment (CEQ) is any item of equipment capable of communicating across an ATM network. In the future, a wide variety of equipment is likely to be available which is capable of using the broad bandwidth and other full benefits of ATM. One of the most popular of today's visions is the concept of *multimedia* applications – devices capable of enabling their users simultaneously to transmit synchronized video, electronic mail, data applications and telephone messages over the same line at the same time.

In the short term, ATM user equipment is likely to take the form of a *broadband terminal adaptor (B-TA)*, allowing existing devices to communicate using an ATM network. Some of these types of devices are already available, with various commercial names (e.g. ATM PAD).

Customer equipments communicate with one another across an ATM network by means of a *virtual channel*. This may either be setup and cleared down on a call-by-call basis similar to a telephone network, in which case the connection is a *switched virtual circuit (SVC)* or it may be a permanently dedicated connection (like a leaseline or private wire), in which case it is a *permanent virtual circuit (PVC)*.

ATM Multiplexor

An ATM multiplexor is usually intended to allow different *virtual channels* from different ATM UNI ports to be bundled for carriage over the same physical transmission line. Thus two or three customers outlying from the main exchange (Figure 2.6) could share a common line. Returning to our analogy with the postal system, the ATM multiplexor performs a similar function to a postal sack – it makes easier the task of carrying a number of different messages to the sorting station (ATM switch) by bundling a number of *virtual channels* into a single container, a *virtual path.* There is more about virtual channels and virtual paths later in the chapter (Figures 2.9–2.11).

ATM crossconnect

An ATM crossconnect is a slightly more complicated device again than the ATM multiplexor. It is analogous to a postal depot, where the various sacks of mail are unloaded, sorted and adjusted into different van loads. As at the postal depot, where the sacks of mail remain unopened, so the usual function of an ATM crossconnect (the cross-connection of *virtual paths*) leaves VP contents, the individual *virtual channels*, undisturbed. The ATM crossconnect appears again in Figure 2.11.

ATM switch

A full ATM switch is the most complex and powerful of the elements making up an ATM network. It is capable not only of cross connecting *virtual paths*, but also of sorting and switching their contents, the *virtual channels* (Figure 2.12). It is the equivalent of a full postal sorting office, where sacks can either pass through unopened, or be emptied and each letter individually resorted. It is the only type of ATM node device capable of interpreting and reacting upon user or network signalling for the establishment of new connections or the clearing of existing connections.

ATM Network Node Interface (NNI)

The ATM *Network Node Interface (NNI)* is the interface used between nodes within the network or between different subnetworks. A standardized NNI gives the scope to build an ATM network from individual nodes supplied by different manufacturers. Alternatively, it can be used to provide a standardized interface between ATM subnetworks provided by different manufacturers (e.g. IBM *Nways* and *Stratacom*). In this case NNI is sometimes referred to as *network–network interface.*

ATM Inter-Network Interface (INI)

The *inter-network interface (INI)* allows not only for intercommunication, but also for clean operational and administrative boundaries between interconnected ATM networks. It is based upon the NNI but includes more features for ensuring security, control and proper administration of inter-carrier connections (i.e. where networks of two different operators are interconnected). ATM forum calls this interface *B-ICI (broadband inter-carrier interface).*

2.6 The Types of Connections Supported by an ATM Network, and the ATM Adaptation Layer

ATM is a *connection-oriented* method of creating telecommunications paths across a switched network. This means that a mechanism exists for setting up a connection between two points of the network for the duration of a call. A *connectionless service* would be one in which sender and receiver were not simultaneously connected to the network. The postal service is like a connectionless service, since the sender posts his letter, and will be doing something else at the time the receiver receives it. Had the receiver first been forewarned that a message would be sent (the equivalent of call setup) and then had to stand ready at his postbox until he received it, while the sender stood by the postbox until he received confirmation of receipt, then the postal service would be a connection-oriented service.

The connections created by an ATM network provide a flexible communications medium for all types of telecommunications services, including not only *connection-oriented* services but also *connectionless* services. These services are defined in ITU-T recommendations F.811 (broadband connection-oriented bearer service, BCOB) and F.812 (broadband connectionless data bearer service, BCDBS).

An extra functionality is added to a basic ATM network (correctly called the *ATM Layer*) to accommodate the various different types of connection-oriented and connectionless network services. This functionality is contained in the *ATM adaptation layer*. The *ATM adaptation layer (AAL)* lays out a set of rules how the 48-byte *cell payload* can be used, and how it should be coded.

Table 2.1 Service classification of the ATM Adaptation Layer (AAL)

Transmission Characteristic	Class A	Class B	Class C	Class D
AAL Type	AAL Type 1 (AAL1)	AAL Type 2 (AAL2)	AAL Type 3/4 (AAL3/4) AAL Type 5 (AAL5)	AAL Type 3/4 (AAL3/4) AAL Type 5 (AAL5)
Timing relation between source and destination	Required	Required	Not required	Not required
Bi rate	Constant	Variable	Variable	Variable
Connection mode	Connection-oriented	Connection-oriented	Connection-oriented	Connectionless

These special codings enable the end-devices which are communicating across the *ATM Layer* to support optimally the desired connection-oriented or connectionless service as needed.

The services offered by the *ATM adaptation layer (AAL)* are classified into four *classes* or *types* (the standards use both terminologies). The distinguishing parameters of the various classes are listed in Table 2.1.

An example of a Class A service is circuit emulation (i.e. clear channel connections like hard-wired digital circuits). In the ATM specifications such services are referred to as *constant bit rate (CBR)* or *circuit emulation services (CES)*. Thus a constant bit rate video or speech signal would be an AAL Class A service and would use AAL1.

Variable bit rate (VBR) video and audio is an example of a class B service. Thus an audio speech signal which sends no information during silent periods is an example of a Class B VBR service and would use AAL2.

Class C and Class D cover the connection-oriented and connectionless data transfer services. Thus an X.25 packet switching service would be supported by a Class C service, and connectionless data services like electronic mail and certain types of LAN router service would be Class D. Both classes C and D use AAL types AAL3/4 or AAL5.

2.7 The Concept of Virtual Channels and Virtual Paths

Being *connection-oriented*, ATM networks establish connections for the purpose of communication between network end-users. Actually, the end-to-end connection established between two CEQ (customer equipments) communicating across an ATM network (the *ATM Layer*) is correctly called a *channel.*

Unlike telephone and other *circuit-switched* networks, ATM Layer *channels* take the form neither of separate physical wires nor of dedicated signal bandwidth. Instead the channels are achieved by statistical multiplexing, whereby a number of channels actually share the same physical connection and bandwidth. The channels nonetheless appear to their users to be independent of one another. Since the channels appear to be separate, even though they share the same physical connection, they are called *virtual channels* (or sometimes *logical channels*).

A *VC* or *virtual channel* is thus the connection established between two customer equipments (CEQ) communicating across an ATM network (at the *ATM Layer*).

At this point, the terminology may appear to get even more confusing, but Figure 2.7 should help us to understand the further vocabulary defined in the next part of our discussion, and also bring more meaning to the ATM multiplexors, crossconnects and switches we have already briefly introduced.

A *virtual channel* extended all the way across an ATM network (*ATM Layer*) is actually a *virtual channel connection (VCC)*. This connection may be composed of a number of shorter length *virtual channel links*, which when laid end-to-end make up the VCC.

A *virtual channel link* is a part of the overall VCC, and shares the same endpoints as a *virtual path connection (VPC)*. The idea of a *virtual path (VP)* is valuable in the overall design of ATM networks. As we have already discovered in the earlier part of the chapter, a *virtual path* has a function rather

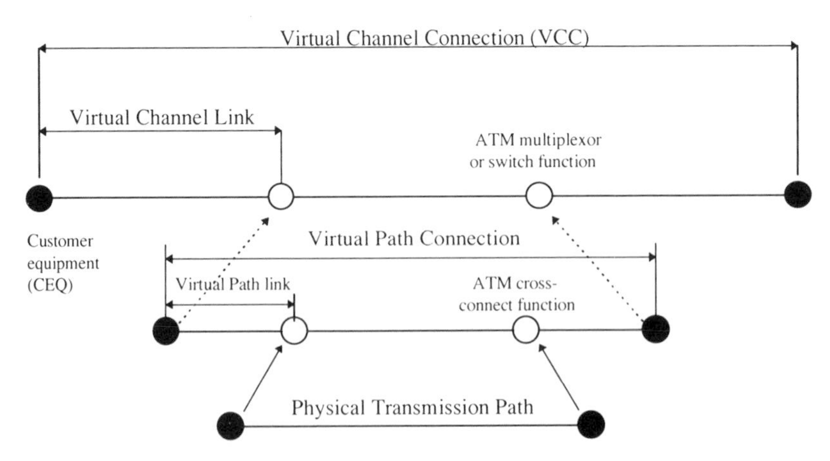

Figure 2.7 The relationship between Virtual Channels, Virtual Paths and Physical Transmission Paths

like a postal sack. In the same way that a postal sack helps to ease the handling of letters which all share a similar destination, so a *virtual path* helps to ease the workload of the ATM network nodes by enabling them to handle bundled groups of *virtual channels.* Thus a *virtual path (VP)* carries a number of different *virtual channel links*, which in their own separate ways may be concatenated with other virtual channel links to make VCCs.

Just like *virtual channels, virtual paths* can be classified into *virtual path connections (VPCs)* and *virtual path links,* where a VPC is made up by the concatenation of one or more virtual path links. A *virtual path link* is derived directly from a physical transmission path.

All virtual channels in an ATM network (be they VCCs or VPCs) can be classified according to whether they are *permanent* or *switched virtual circuits (PVC* or *SVC).* A *permanent virtual circuit (PVC)* can usually only be established by a human network operator by manual command. A *PVC*, like a leaseline, is a circuit permanently connected between the same two (or more) end-points. An *SVC (switched virtual circuit)* is analogous to a telephone call connection. A new SVC is set up with each connection request (new call) to the destination requested in the dialling (or *address*) information.

2.8 How is a Virtual Channel Connection (VCC) Set Up?

One or more *virtual channel connections* or *virtual path connections* within a UNI or NNI link may be allocated for signalling, either user-to-network signalling (at the UNI) or node-to-node signalling (at the NNI). These connections are then called *signalling virtual channels* (*SVC*, but not to be confused with *SVC – switched virtual connection*).

Information about all the other *virtual connections* available between the user and the node at the UNI, or between the two nodes at the NNI, is carried by the *signalling virtual channel.* Thus a signalling message sent over the *SVC (signalling VC)* might be 'set up virtual connection number one between user A and user B'. Another message might be 'clear the connection between A and B'. Unlike a telephone connection, where the dialled digits and other control information are transmitted over the channel itself (*channel associated signalling*), ATM instead uses a single, dedicated channel for all the signalling information (i.e. uses *common channel signalling*). Drawing the parallel between ATM signalling and narrowband ISDN signalling, the UNI signalling interface is equivalent to the ISDN signalling defined by ITU-T recommendation Q.931 (and indeed is based upon it and specified in Q.2931), and ATM NNI signalling is equivalent *to signalling system 7* as used in ISDN.

PVCs (permanent virtual circuits) are set up by a human network operator.

They are connections which are hard-defined (by network 'configuration') always to exist between the same two (or more) endpoints. For their setup, either the user could telephone the network operator or a signalling virtual channel could be used to carry an equivalent 'electronic mail' message to the network management centre operator. The human operator would then issue the relevant network management commands, which would be relayed by signalling virtual channels between network nodes to instruct the various network devices to establish the connection (e.g. assigning VCI, VPI as we will discover in the next section of this chapter). PVCs remain in place as permanent connections until deleted or reconfigured.

When setting up an SVC (switched virtual circuit) *connection*, a negotiation is first conducted between the *customer equipment (CEQ)* and the network over the *UNI signalling virtual channel*. The negotiation establishes the endpoints of the desired connection, required technical characteristics (e.g. connection type and bitrate, etc.), and the necessary quality and priority of the connection (see section 2.12 later in this chapter). During the negotiation, virtual paths and connections between the various nodes and other equipments are allocated, and the reference numbers of these connections, the combination of *virtual path identifiers (VPIs) and virtual connection identifiers (VCIs)*, are confirmed over the signalling channel. These values (VPI and VCI) then appear in the *header* of any cells sent, in order to identify all those cells which relate to this connection.

The functionality or device which exists at the end of a signalling virtual channel and conducts the act of signalling is called a *signalling point (SP)*. Such functionality exists in customer equipment (CEQ) and in ATM switches. A *signalling transfer point (STP)* is a switching point for the information carried in signalling virtual channels. Using a single SVC (signalling VC) connection via an STP, an SP may communicate signalling messages to numerous other SPs. Typically STPs are used (as with signalling system 7) to improve the efficiency and reliability of the signalling network (see Figure 2.8).

In Figure 2.8, Customer Equipments (CEQs) A and B are to communicate

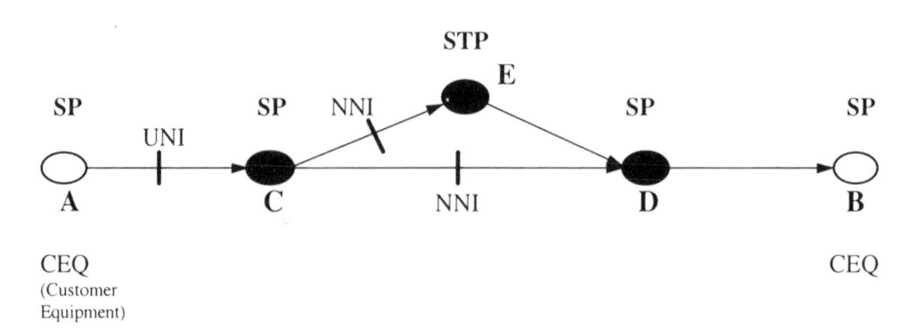

Figure 2.8 Signalling Points (SPs) and Signalling Transfer Points (STPs)

over ATM network nodes C, D and E. The signalling point (SP) in A signals the connection request to the SP in node C over the UNI signalling virtual channel. Node C decides that the connection itself should be set up direct from node C via node D to node B. Signalling information therefore needs to be sent to the SP at node D. This can either be done using a directly established signalling virtual channel between nodes CD, or as in our case, this signalling virtual channel (and its expensive terminating equipment) can be avoided by instead sending the signalling information between SPs C and D via the STP at E. In this way, the SP at node C could also communicate with other nodes in the network via a single signalling virtual channel (that to the STP at node E).

2.9 *Virtual Channel Identifiers (VCIs)* and *Virtual Path Identifiers (VPIs)*

As we have seen, connections are set up in ATM networks as *virtual channel connections*. These in turn are comprised of shorter sections called *virtual path connections*, which are concatenated end to end. The various virtual channels and virtual paths are *identified* by reference numbers carried by the cell headers of active connections. The identifying reference numbers are called *virtual channel identifiers (VCIs)* and *virtual path identifiers (VPIs)*.

Figure 2.9 illustrates how a physical connection may be subdivided into a number of different virtual paths, each with a unique VPI. Each of these virtual paths may then be subdivided into a number of separate virtual channels, each with a separate VCI. These are then the virtual channels which can be concatenated (by means of an ATM multiplexor, crossconnect or switch) to make up an end-to-end connection between CEQs.

By understanding how VPIs and VCIs are assigned and processed, we can understand much better how multiplexors, crossconnects and switches operate in an ATM network.

In order to specify a particular virtual channel within a particular physical

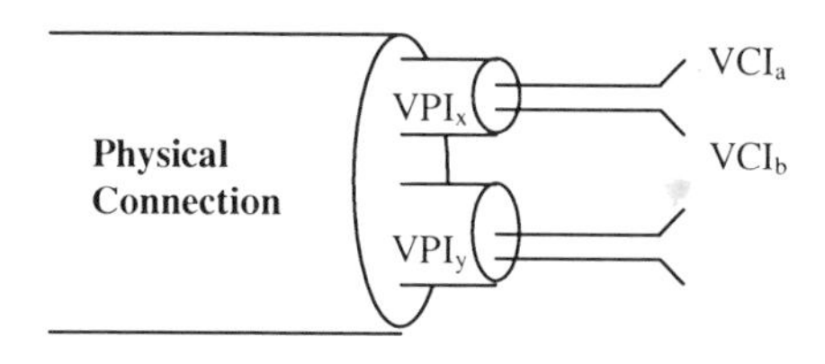

Figure 2.9 Virtual Path and Virtual Channel Identifiers

connection, both the VPI and VCI must be stated. This is because while the VPIs are unique for each interface (i.e. each individual UNI or NNI), the VCIs are not unique. For each virtual path, the numbering of virtual channels may begin at '1'. The VPI/VCI combination, however, is unique and is sufficient to identify any active connection at the interface (i.e. on the same physical connection).

As we have already seen, the function of ATM multiplexors, crossconnects and switches is to concatenate the various channel sections, to switch VCs and to crossconnect VPs. Figures 2.10–2.12 illustrate in detail the normal functions of each of these types of devices.

An ATM multiplexor (Figure 2.10) allows a number of virtual channels from separate virtual paths to be combined over a single virtual path. Thus the virtual channels carried by physical links 1, 2 and 3 of Figure 2.10 are combined together into a single virtual path carried by physical link 4. Thus in Figure 10, three separate end-user devices use separate virtual channels (with different VCIs) to be connected to a remote ATM switch by means of a single physical connection line and an ATM multiplexor.

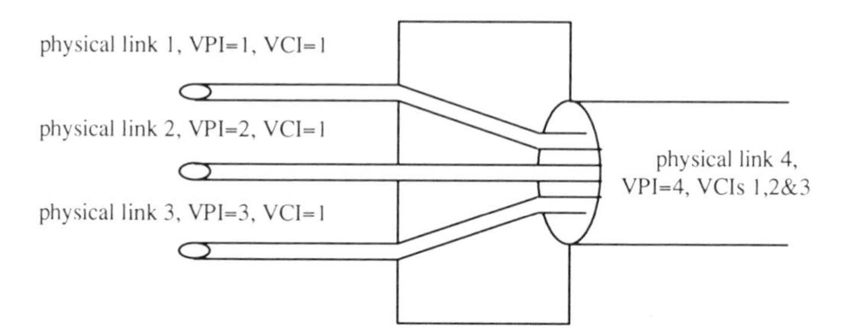

Figure 2.10 ATM multiplexor

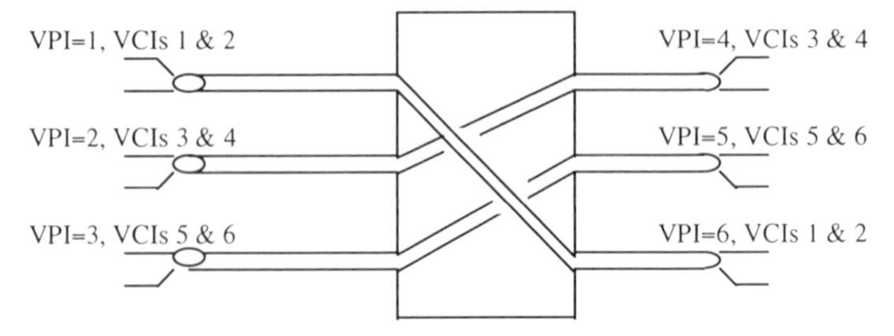

Figure 2.11 ATM crossconnect

An ATM crossconnect (Figure 2.11) allows rearrangement of virtual paths without disturbance of the virtual channels which they contain. Thus in Figure 2.11, the contents of virtual path VPI = 1 are crossconnected to virtual path VPI = 6; the VCIs remain unchanged. In effect, an ATM crossconnect is a simple form of ATM node, which need only process (and *translate*, i.e. amend) the VPI held in the cell header. It is potentially faster-running and cheaper than a full-blown ATM switch.

A full ATM switch (Figure 2.12) must have the capability not only to crossconnect virtual paths, translating as necessary VPIs held in cell headers, but also to switch virtual channels between different virtual paths. This requires the additional ability to process and translate VCIs held in cell headers. It is thus a relatively complex device. Good performance depends upon very fast processing of both VPIs and VCIs in the cell headers. A full ATM switch is consequently a costly device.

2.10 Information Content and Format of the ATM Cell Header

The main function of the cell header is to carry the VPI and VCI information which allows the active network elements (ATM multiplexors, crossconnectors and switches) to switch the cells of active connections through the network. The exact format of the cell header is shown in Figure 2.13.

As is clear from Figure 2.13, the cell header comprises 40 bits, of which 24 (UNI) or 28 (NNI) are used for the virtual path and virtual channel identifiers. Together, these (the VPI / VCI fields) are called the *routing field.* There are four other *fields* which occupy the remainder of the header. The *PT (payload*

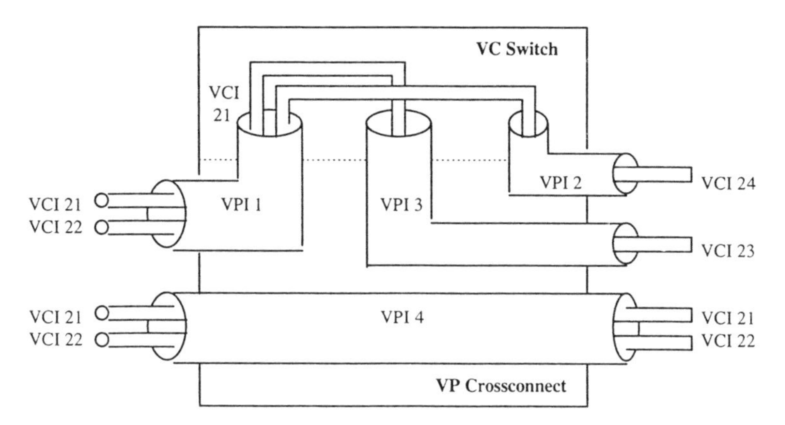

Figure 2.12 Full ATM switch

8	7	6	5	4	3	2	1	Byte Octet
GFC (at UNI), **VPI** (at NNI)				VPI				1
VPI				VCI				2
VCI								3
VCI				PT			CLP	4
HEC								5

GFC = Generic Flow Control
VPI = Virtual Path Identifeir
VCI = Virtual Channel Identifier
PT = Payload Type
CLP = Cell Loss Priority
HEC = Header Error Control

Note: the GFC is used only at the UNI, as the NNI bits 5–8 of octet 1 are used as VPI.

Figure 2.13 Structure of the ATM cell header

type) field is occupied by the *payload type identifier (PTI)*. This identifies the contents of the cell (the *information field* or *payload*) as either a user data cell, a cell containing network management information, or a resource management cell. The *cell loss priority (CLP)* bit, when set to value 1 means that the cell should be discarded prior to cells where the *CLP* value is set at 0.

The *generic flow control (GFC)* field is used to control the cell transmission between the customer equipment (CEQ) and the network. The GFC field is used to alleviate congestion at ATM multiplexors, where a number of CEQ devices are sharing the same trunk line connection to the network (Figure 2.14).

When there is no trunk congestion (i.e. there is no appreciable accumulation of cells waiting in the multiplexor buffer to be transmitted over the trunk) then the GFC field is set to the *uncontrolled transmission mode*. In this mode each of the CEQ devices may send cells to the network (i.e. to the multiplexor for transmission over the trunk). All bits are set to zero in the uncontrolled mode.

However, if there is a sudden surge of cells from all of the CEQ devices and the multiplexor experiences congestion (i.e. the filling of its cell buffers to a critical threshold level) then the GFC field is used to subject the cell flow from the various CEQ devices to *controlled transmission*. This limits the rate at

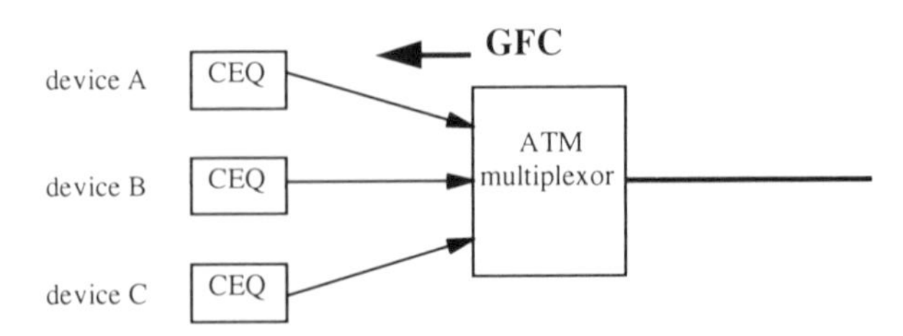

Figure 2.14 Generic flow control regulates trunk congestion at a multiplexor

which the CEQ devices may continue to send cells of one or more different types to the network.

The *generic flow control* procedure is an asymmetric procedure used to prevent originating CEQ devices from overloading the network. The CEQ device is expected to accept or deal with all cells sent to it, without means for slowing the rate of delivery from the network.

GFC may be a useful and important technique, but it has only limited significance – where standardized local congestion relief functions are defined and supported both in the network and in the customer equipment.

The *header error control (HEC)* field comprises 8 bits used for detection of errors in the cell header caused during cell transmission. The function can be set in either *correction* or *detection mode*. In correction mode, single bit errors may be detected and corrected. In detection mode, one or multiple errors may be detected and the affected cells discarded.

The information held in the header is transmitted in the order of octets (i.e. octet 1 first, followed by octet 2, then octet 3, etc.). The *most significant bit (MSB)* of each octet (i.e. bit 8) is transmitted first. The payload is similarly transmitted in the order of octets, MSB first.

2.11 The Basic Architecture of ATM

The ATM standard specifications refer to the various layers which go to make up an ATM network. The most important of these are set out in Table 2.2.

The physical layer defines the physical transmission types which are suitable for ATM. In particular the specifications are so designed to make *SDH (synchronous digital hierarchy)* transmission lines suitable. *SDH* is the most modern form of telecommunications transmission technology. The *physical*

Table 2.2 Hierarchical architecture of an ATM network

Higher level communication	Higher layer protocols		
ATM network	ATM Adaptation Layer (AAL)		
	ATM transport network	ATM layer	VC level VP level
		Physical Layer	Transmission path level Digital section level Regenerator section level

layer specifications define the electrical, optical and transmission characteristics which should be used, as well as the interface required by the *ATM Layer*.

The *ATM layer* is the next higher functionality added to the physical transmission. Its specifications define the 48-octet cell, the 5-octet header, and the methodology to be used when transmitting cells across the physical network.

The *ATM adaptation layer (AAL)* specifications define how cells may be used to create connections suitable for a wide range of end-uses (e.g. *constant bit rate (CBR)* connections, voiceband signal transport, data transport, etc.). The various types of connections are classified by the AAL specifications into the five classes already discussed in Table 2.1.

The *higher layer communication* is the useful information carried between the end-terminals by the ATM network.

2.12 Traffic Control and Congestion Control in an ATM Network

Despite being based on statistical multiplexing techniques, ATM networks are designed to meet very high standards of transmission. This is achieved by powerful traffic control, congestion control and quality of service functions built into the ATM signalling mechanisms and flow control techniques.

At the time when a CEQ (customer equipment) requests to set up a new connection across an ATM network, it must first negotiate with the network for the connection, declaring the required *peak cell rate*, *quality of service (QOS) class* and other connection type parameters needed. The *connection admission control (CAC)* function at the first ATM signalling point (ATM exchange) then decides whether sufficient resources are available to allow immediate connection. If so, the connection is set up. If not, the connection request is rejected in order to protect the quality of existing connections. (The analogy is the telephone user's receipt of busy tone when no more lines are available.)

The quality of service offered by an ATM connection depends upon the number of cells lost or discarded during transmission. Cells from low-priority connections are discarded first at times of network congestion in order to help maintain the quality of more critical high-priority connections.

The *transfer capacity* is the maximum number of bits per second which may be transferred by a particular path or transmission link. An SDH *STM-1* type transmission line, for example, has a transfer capacity (or *transfer capability*) of 155 520 kbit/s. In contrast, the *throughput* of an ATM connection is the number of user data bits (i.e. *payload*) successfully transferred across the network per second. The throughput is always significantly less than the *transfer capacity*, due to idle periods and network *overheads*.

3

Why Do We Need ATM?

Important benefits of ATM are its abilities for broadband switching, for fast data switching, for variable bandwidth switching and for supporting integrated networks in multimedia. In this chapter we explain why these qualities have come about. We explain the factors which have influenced ATM development, why it is so important, and some of the ways in which we will use it.

3.1 The Drive towards Broadband Networks (*B-ISDN*)

The development of ATM has been characterized by a number of significant breakthroughs, each caused by a different development group recognizing a different potential in the technology.

Much of the initial development started in the data networking field, as engineers tried to develop *fast packet switching* techniques suitable for the carriage of emerging video and so-called 'image' *applications*.

The high bandwidth (i.e. bitspeed) demanded by modern computer window-based and graphics software and emerging moving image or even interactive video *applications* has long been beyond the technical capabilities of classical *wide area* packet data networks (e.g. X.25 or SNA). Furthermore, while the use of dedicated high-speed leaselines or multiplexor networks (e.g. TDM multiplexors) may provide a technically adequate solution for providing the necessary bandwidth, the economics (i.e. the high costs of long distance lines) have limited high-speed applications to in-building or on-campus use.

The demands of new graphical computer software, the emergence of videotelephony and cable television, and the political pressure for the development of the *information highway* have all combined in their various ways to stimulate the development of switched *broadband* networks. Quite simply, what was needed were networks capable of delivering high bandwidths to their users on a switched basis – the switching providing an economic means of cost sharing. The result of the pressure for switched broadband networks has been the development of *broadband ISDN (broadband integrated services digital network*, or *B-ISDN*).

Under the banner B-ISDN a number of the world's technical standards development committees (including the International Telecommunications Union (ITU-T), ETSI (European Telecommunications Standards Institute) and ANSI (American National Standards Institute)) have developed specifications, but the work is far from complete. To date the specifications describe the general functions which will be performed by a B-ISDN, and some of the network interfaces, so that the concept of B-ISDN to date is more *what* will be achieved than the full details as yet of exactly *how*.

The current most advanced solution for B-ISDN is based on ATM, but this does not preclude other realizations, due either to the technical unsuitability of ATM for some requirements or to the established technical interests of particular network equipment manufacturers.

Figure 3.1 illustrates the switched broadband capabilities of B-ISDN (broadband ISDN). The two main capabilities are for:

- switching of high bandwidth connections across the network, thereby enabling the sharing of long distance, high bandwidth connections between many users;

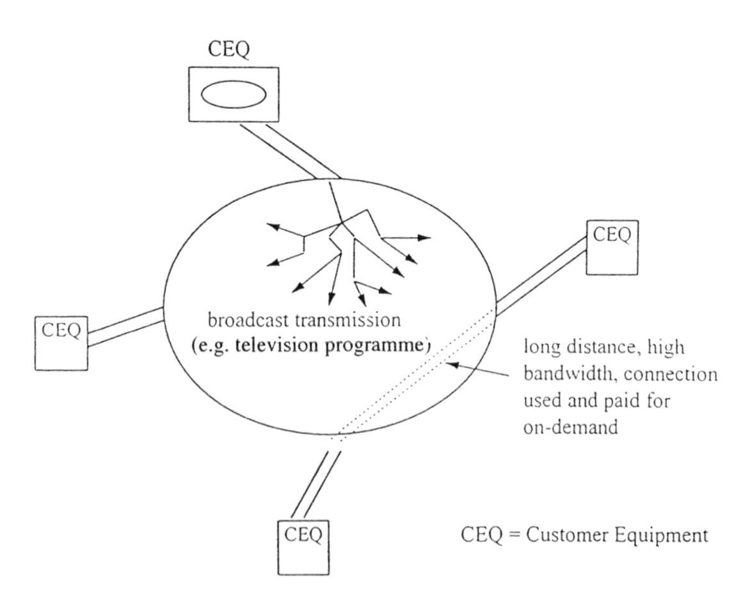

Figure 3.1 The basic switching capabilities of B-ISDN

- broadcast distribution of high bandwidth signals (e.g. video or television programmes).

In the ATM realization of B-ISDN, different types of connections (e.g. data connections, video connections, *connection-oriented* connections, *connectionless* connections etc.) are available, each type tuned to the needs of a particular application type. Switched video connections, for example, will be valuable for users of high quality picture or video telephones. The broadcast capability, meanwhile, will enable cable TV companies to develop economically efficient metropolitan fibre cable networks for the distribution of pay-TV or *video-on-demand*.

3.2 The Need for Fast Responding Data Networks

While the basic need to carry high bandwidth signals has driven the need for the development of broadband networks, so has the need for faster responding networks. It may not be immediately obvious, but the time required to propagate even low bandwidth information across a digital network is dependent upon the bitspeed employed – the higher the bitspeed, the lower the propagation delay. Put more simply, even data applications with limited information transport needs appear to run faster when carried by a broadband network, even though sufficient bandwidth may already have been available previously. The following discussion explains two reasons why.

Let us imagine two rainwater conduits, one of small bore and one of large bore. Let us assume that the first has throughput capacity of 5 litres/s and the second of 10 litres/s. Now let us assume that the rainfall rate is 4 litres/s. Why should I bother with the large bore conduit? The answer is that the rainfall rate is not constant. Over the course of time the rate may vary between, say, 2 and 6 litres per second, so that during moments in time when the rate of rainfall exceeds 5 litres per second, water will be accumulating in the roof gutter rather than flowing down the conduit. The accumulation clears when the rainfall rate drops momentarily below 5 litres/s. As a result of the momentary accumulation, some of the rainwater is delayed slightly – thereby increasing the 'propagation delay'. The analogy is equally relevant to data transmission, where the rate of generation of typed characters or other data to be transferred is not constant, but can wildly fluctuate.

Figure 3.2 illustrates a more detailed example of data transmission across a telecommunications transmission line. In this case, the carriage of the electrical signals (i.e. the waveform pulses representing individual *bits* of a digital signal pattern) is close to the speed of light. Thus the leading edge of an individual pulse traverses the network at around 10^8 m/s (Figure 3.2(a)). The

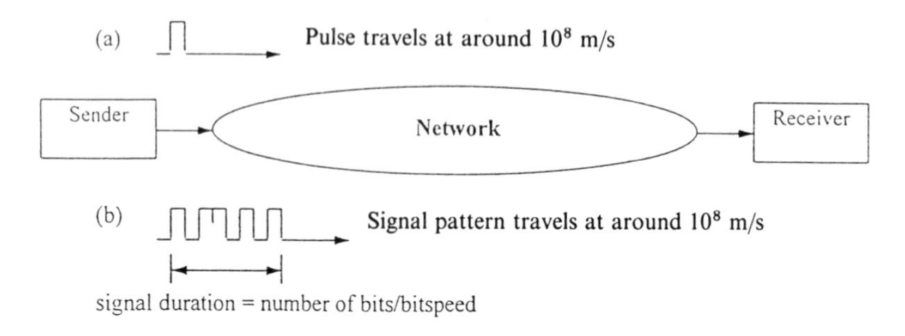

Figure 3.2 Signal propagation time across a data network

time required, however, to transmit an entire byte (8 bits) is sensitive to the transmission bitspeed. The propagation time in this case is equal to the sum of the raw propagation time and the signal duration (Figure 3.2(b)).

Taking the example of a 9600 bit/s dataline of 100 km length (as might be employed in a corporate packet switching network today), we can calculate propagation times for both cases (a) and (b) of our example:

- Single bit (pulse) propagation time $= 10^5\ \text{m}/10^8\ \text{m s}^{-1} = 10^{-3}\ \text{s} = 1\ \text{ms}$
- Byte propagation time = pulse propagation time + signal duration
  $= 1\ \text{ms} + 8\ \text{bits}/9600\ \text{bit s}^{-1}$
  $= 1.8\ \text{ms}$

In other words, before enough bits (8) have been received to interpret the data character (e.g. ASCII character) being transmitted, 1.8 ms have elapsed. This compares with the 1 ms needed for conveyance of a pulse across the line. Despite the fact that the average throughput required from the line may be far less than the 9600 bit/s available (say 2400 bit/s or 300 characters/sec), the effective propagation time of characters across the line is much longer than the 1 ms which you might expect.

No human will notice the extra 0.8 ms, you might say? Indeed they won't where a simple one-way transmission is involved with a human end user. But where an interactive dialogue is taking place between two computers (question-answer-question-answer), then this will take around 80% longer to conduct. A human waiting for the computer's response may see a response in around 4 seconds where previously it was around 2 seconds. Such intercomputer dialogues are the main cause of delays for modern computer software. (Typical dialogues run 'please send first character' – *'first character'* – 'received first character OK, please send next character' – *'second character'* – 'received second character OK...' etc.)

The perhaps surprising reality is that it may indeed make sense to use a network with 64 kbit/s transmission links rather than 9600 bit/s links even though the average throughput is only 4000 bit/s. (Reduction in byte propagation time from 1.8 ms to 1.1 ms). The following points are thus critical in the response time performance of data networks and associated computer applications software:

- transmission line bitspeed (e.g. 9600 bit/s, 64 kbit/s, 2 Mbit/s, 155 Mbit/s etc);
- message or packet length in number of bits;
- the number of inter-computer interactions (request and response dialogue) necessary to complete an action before responding to the human user.

Though our example is of a lowspeed data application, similar principles apply for all types of data applications. Thus the higher the bitspeeds employed in the network, the faster the application response time.

The recent explosion in the number of *LANs (local area networks* – networks connecting personal computers within office buidlings) and the need to interconnect LANs, as well as the growing number of *client/server* computing (UNIX) environments have created the demand for broadband networking – from rates starting at 64 kbit/s and extending to over 100 Mbit/s. The result of this demand has been the development of a number of different techniques including *frame relay, FDDI (fibre distributed data interface), DQDB (dual queue dual bus), SMDS (switched multimegabit data service)* and *ATM*. As we discuss in the next chapter, ATM looks likely to become the prevalent technique.

3.3 The Desire for Integrated Voice, Data and Video networks

The early telecommunications networks (i.e. telegraph, telephone and telex) were all *circuit-oriented*. A physical connection was established between caller and called party for their exclusive use for the duration of the call. The physical connection could then be used to carry any type of signal, as required. The advent of modems made possible the first data communication across a telephone network. In effect we had the world's first *integrated network* – capable of carrying both voice or data.

But, prompted by the rapid growth of the data processing industry in the 1970s, a new specialized industry emerged to cater for the needs of data communications. Separate telephone and data networks emerged, based on entirely different switching techniques. The data switching technique made possible much more efficient carriage of data than was possible using modems and a switched telephone network. It was based on statistical multiplexing, as we discussed in chapter 2.

Ever since the emergence of separate telephone and data networks, engineers have been seeking to develop *integrated networks*. The basic hypothesis of such networks is simple – that there is significant potential for savings in line costs by the use of a single network infrastructure for the carriage of all types of signals (voice, data, video, etc.).

Figure 3.3 illustrates the basic philosophy of integrated networks. Figure 3.3(a) illustrates two independent dedicated networks for telephone (voice) and data. Such a realization leads to duplication of the main network backbone, with the creation of two separate networks to cover the same geographical area. By contrast, Figure 3.3(b) illustrates an integrated network serving the same access lines. In this case, both data and telephone equipment is connected to the same network, and shares the same infrastructure. As a result only one network is necessary.

The integrated network has the potential to be more efficient than the combination of dedicated networks. This is because the capacity which might otherwise have lain idle at a particular point in time in the telephone network, may be used to carry data traffic. (Maybe the data traffic is more prevalent at a particular time of day when telephone traffic is slack.) In a similar way, idle data capacity can be used for telephone use. Potential savings arise both within the backbone network and on the access lines. Thus we see in Figure

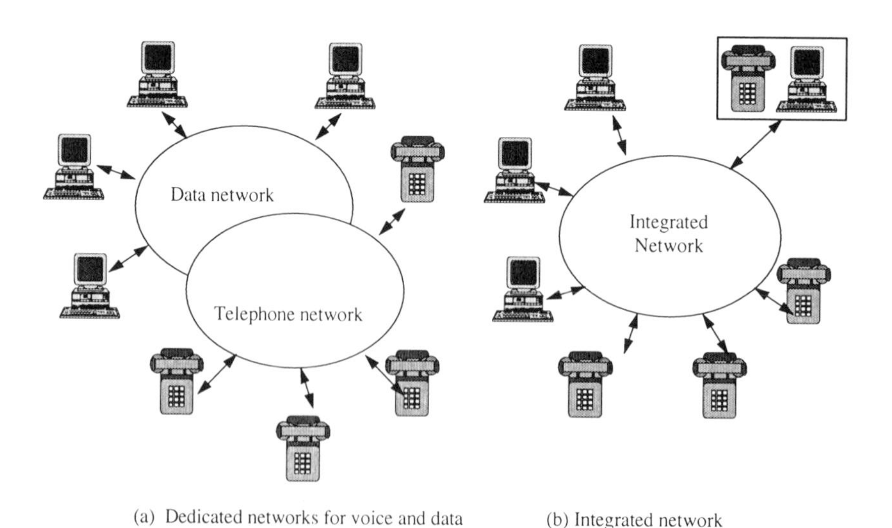

Figure 3.3 The principle of integrated networks

3.3(b) how an access line is saved by the sharing of a line by both data and voice equipment at one of the sites.

The savings due to integration are sometimes outweighed by the losses resulting from the inefficient technical manner in which either the data or the voice is carried. Thus, for example, in an ISDN network, the idle capacity in what is in effect an advanced telephone network is made available for data usage. Unfortunately, the capacity taken up (e.g. 64 kbit/s channel) to carry a given data signal (typically 9600 bit/s) is six times as great as the capacity that would have been needed on a dedicated data network, so that the cost of converting data equipment to ISDN may not be justified by the meagre savings.

The cell relay technique employed by ATM is a data switching methodology which has been adapted to be capable of switching voice. The philosophy of using data switching for voice (rather than voice switching for data) offers a much greater promise for effective and efficient integrated networks, since data networks adapt better to the needs of connections and applications (e.g. computer or telephone) requiring different bitrates.

The huge number and variety of different types of communication devices which today are expected to communicate with one another are pressing the need for integrated networks, and therefore also for ATM. It is no longer tenable to contemplate special networks for each conceivable type of device.

3.4 The Need for Variable Bandwidth Switching

A highly significant attribute of ATM technology is its ability to switch connections at almost any bandwidth. This adds greatly to the flexibility and efficiency of ATM networks when compared with their forbears. Using the analogy of trucks on a road network, no longer must a customer hire a complete freight truck for a single carton, or two trucks when one carton more than a full load was to be carried. Instead the customer hires a truck tailored in size to the load it is to carry. This attribute has advantage in both a data and a voice or video context.

Unlike digital telephone switches which are only able to switch 64 kbit/s channels, an ATM switch is capable of switching connections of 32 kbit/s, 16 kbit/s, 128 kbit/s and so on. We need such other bitrate connections for the many different types of device appearing today. Even for basic telephony, rates other than 64 kbit/s are already common.

New *compression* techniques have made possible very high quality telephone conversation using only 32 kbit/s, 16 kbit/s, even as low as 8 kbit/s (e.g. digital radiotelephone networks, mobile networks, etc.). Meanwhile, there is also growing demand for connections at rates higher than 64 kbit/s (e.g. 128 kbit/s,

384 kbit/s) for the carriage of picturephone signals and *videoconferencing* (video connection of two or more meeting rooms together for purpose of conducting a conference or meeting). Despite these two trends, nearly all modern telephone networks remain unable to switch connections at other than 64 kbit/s. For users that need lower bitrates, the remaining capacity (to fill to 64 kbit/s) is simply wasted (Figure 3.4(a)). For other users, higher bitrates can only be achieved by splitting up the signal to be carried into 64 kbit/s chunks for separate carriage. The high-speed signal must then be reconstituted (at some considerable technical effort) at the receiving end (Figure 3.4(b)).

In some telephone networks, voice compression is used only on the long distance transmission links. In this case it is possible to use one long distance 64 kbit/s transmission channel to carry two compressed voice connections of 32 kbit/s. Voice compression is conducted after the switching stage, and decompression is undertaken at the receiving end of the transmission line before the next switching stage (Figure 3.5). Using voice compression in this way, savings can be made on the expensive long distance transmission links without complicating matters at the switch by presenting connections at rates other than 64 kbit/s.

In Figure 3.5 there are two trunk connections between exchanges A and B and between exchanges A and C. The trunks between exchanges A and B are long distance trunks (say several hundred kilometres). In order to save money a single 64 kbit/s transmission channel has been provided between exchanges A and B and 2:1 voice compression (i.e. from 64 kbit/s to 32 kbit/s) has been applied in order to make the carriage of the two trunks possible. In contrast between exchanges A and C no compression has been applied.

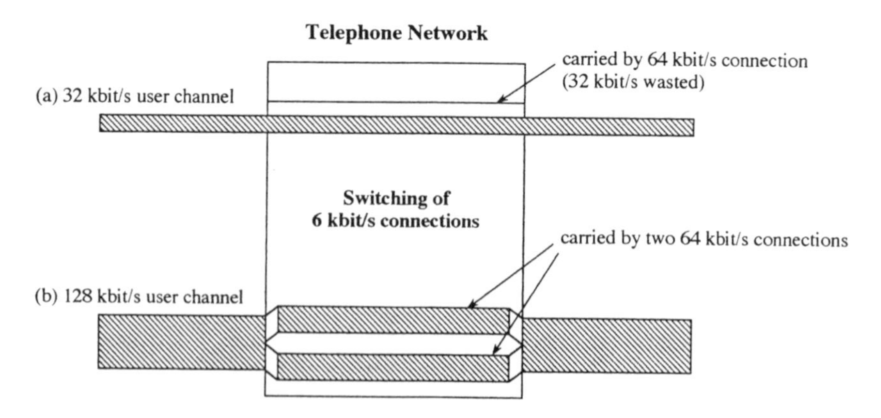

Figure 3.4 Carrying rates other than 64 kbit/s across a digital telephone network

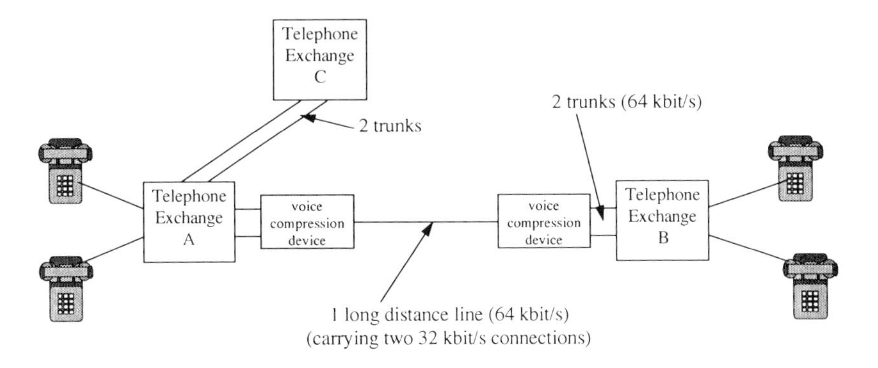

Figure 3.5 Classical use of voice compression in telephone networks

The constellation of Figure 3.5 achieves its main aim – savings on long distance lines without requiring the telephone exchanges to switch other than 64 kbit/s connections. Not only may telephone customers of exchange A talk to customers on exchange B, so can the customers of exchange C. Calls in this case transit exchange A (C–A–B).

A problem arises when the network planner also wants to use voice compression for the trunks between exchanges A and C. The problem is the concatenation of compressed links. The quality of an end-to-end connection from exchange C to exchange B (via A) is probably no longer acceptable. Where such multiple compression–decompression–compression–decompression is undertaken, the signal quality deteriorates rapidly.

What is the underlying problem? The problem is that exchange A is unable to switch connections of other than 64 kbit/s. Only for this reason must we undertake the intermediate decompression and recompression at site A.

One way of getting around the problem would be to leave the signal at 32 kbit/s, switching the transit connections (C–A–B) across exchange A itself using a 64 kbit/s switch path in which half the capacity is wasted (as Figure 3.4 – where a special packing/unpacking device allows the two trunks on either side of the exchange to run at 32 kbit/s while the switch itself switches only 64 kbit/s paths). We could do this, but to do so we not only need to invent the packing/unpacking device, we also create a new problem. Now the customers of exchange A cannot talk with those of exchange B, since there is no longer a compression device available for calls originated at exchange A. So, another dodge, we move the compression to the front side of exchange A, so that all customer calls are compressed to 32 kbit/s before switching. This

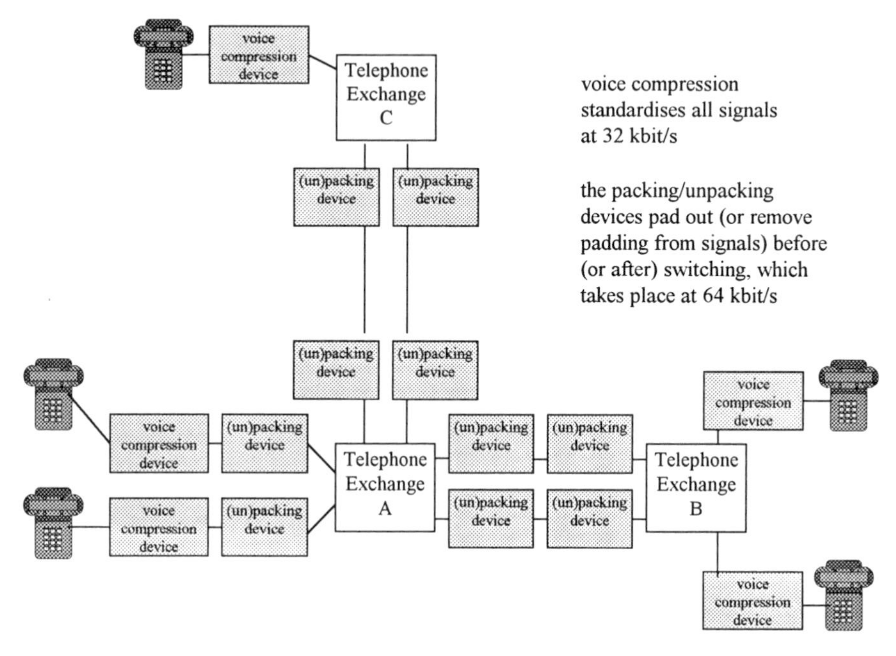

Figure 3.6 A telephone network avoiding multiple voice compression/decompression?

would work. So now we only need get on with designing the device capable of 'packing' the 32 kbit/s compressed signals up to 64 kbit/s for switching through the exchanges and for 'unpacking' them for transmitting to the long distance lines. Bingo – we have the world's first 32 kbit/s telephone network! But it's rather a complicated and confusing mess (see Figure 3.6). It would have been simpler to start from scratch, building a 32 kbit/s switch and telephone network. All well and good as long as 32 kbit/s is the standard bitrate for telephone calls, but when yet another new rate emerges (e.g. 16 kbit/s) then our 32 kbit/s network will turn out to be just as inflexible as its 64 kbit/s predecessor.

A much better philosophy of telephone network design will be to use networks such as ATM which are capable of various and variable bitrates. Telephone connections can be compressed to any desired rate prior to switching and then carried at this rate across the network. When the standard rate of compression changes, the compression devices can be swapped and the network told to set up lower bitrate connections.

The ability to switch connections of almost any desired bitrate is a major advantage of ATM. This means that the network can deliver an optimum bitrate for almost any application (data, video, image applications, etc.).

3.5 The Emergence of *Multimedia*

The word *multimedia* counts amongst the most modern of telecommunications jargon. It is applied to describe user applications which simultaneously employ different types of telecommunications *sessions*. Thus, for example, a *multimedia* application might be one running on a computer workstation which simultaneously permitted the user (in different *windows* presented on the screen) to send electronic mail and conduct a videoconference with his colleagues, while receiving separately another video signal and a fax. The requirement arises from the emergence in the computing world not only of new powerful video and image-based applications but also of multiple user computer operating systems such as *UNIX*, IBM's *OS/2* and *Windows NT*. In order to support *multimedia* applications, a network should offer a wide range of services (e.g. voice, data, video etc.) over an *integrated* network infrastructure. As one of the best *integrated network* technologies, ATM is bound to determine the future for *multimedia*.

3.6 The Problems of Managing Modern Networks

Modern networks are generally larger than their predecessors and have much more complicated topologies, but like their predecessors still continue to grow. The task of network management has become much more difficult, but fortunately this has stimulated the development of much more advanced network management and administration tools. Compared with their predecessors, modern networks are easier to design, extend, administer and maintain in operation.

Modern *network management systems* typically provide a great deal of graphical information about the status of the network, and are able to respond dynamically to switch or link failures in the topology or to unforeseen usage demand. In addition, when adding further nodes or links to some types of networks, some networks automatically 'learn' about the new topology and can direct future calls accordingly. Thus the new node reports to its neighbour, which in turn reports the new arrival to all the other nodes meanwhile advising the new arrival about the network as a whole. Such characteristics are, for example, common in LAN *router* networks (such as the *Internet*). It would be impossible to load the entire *Internet* network topology into each new node that was added.

Simply because ATM is a modern technology, manufacturers of network equipment are likely to provide more advanced network management. Thus, for example, the *BBNS (broadband network services)* architecture which stands behind the IBM ATM product *Nways (IBM 2220)* includes the full *IBM Netview* network management capabilities and a self-learning topology management system like a LAN router network. Simply because ATM networks are modern, they are likely to be easier to manage.

In addition, ATM networks will be easier to manage because they are optimized for use over *SDH (synchronous digital hierarchy)* transmission networks. These are modern, efficient and easily manageable networks for basic point-to-point transmission. It is not so that ATM gains any particular direct benefit from a synergy with SDH. However, when compared with earlier voice and data technologies which do not pass as efficiently within the SDH structure and are unable to take advantage of the highbit rate containers, ATM technology could be said to be more suited to SDH than earlier technologies. ATM networks may thus count SDH advantages among their benefits. In Appendix 1 we discuss the basic technical principles of SDH. The appendix is intended as a simple guide for those readers to whom it is new, and as a quick reference for others. Understanding SDH is important in the context of ATM networks, since it will often be the choice of transmission system for trunks running at speeds exceeding 155 Mbit/s.

4

How Does ATM Compare?

How do the attributes of ATM compare with existing technological alternatives? When is ATM best and when should an alternative be preferred? In this chapter we provide a lexicon of the various different technological means for interconnecting telecommunicating devices and compare their relative strengths against ATM. Included in our review is a comparison with leaselines and TDM networks, with X.25 packet switching, frame relay, LANs, SMDS (DQDB) and FDDI data networks, and with ISDN and intelligent telephone networks.

4.1 Direct Cabling versus ATM

Direct cabling between communicating devices is only feasible where the devices are within close proximity of one another. Why? Because first, even at relatively low line speeds (e.g. 9600 bit/s lines conforming to *recommendation V.24* – as might be used by a laptop or other PC), the line length cannot technically be much longer than about 20 metres without some form of telecommunications line equipment. (At higher line speeds only very short cable lengths may be possible.) Secondly, on a large building or campus site, the number of cables needed to directly connect all possible devices makes direct cabling impracticable to consider, so that some sort of crossconnect device or active switch (like TDM, LAN, ATM, FDDI, etc.) is necessary.

4.2 Point-to-point Leaselines versus ATM Constant Bit Rate (CBR) or Circuit Emulation Service (CES)

Leaselines are lines rented from the telephone company for telecommunications usage between two fixed endpoints in different geographic locations. In effect they provide a dedicated wire (or *private wire*) between the two end devices. A leaseline is a line conditioned for long distance and uninterrupted communication.

Leaselines (or *private* wires as they used to be called) are the basic transmission means for *private* voice or data networks. By *private network* we mean one which uses its own switching equipment, i.e. operates entirely independently of the public telephone or public data network. Historically, leaselines have been used as the basis for company private voice networks (enabling the direct interconnection of office telephone exchanges (PBXs) in different sites) and for company data networks (typically packet switchbased *X.25* or *SNA (systems network architecture)* networks for connecting terminals in remote sites back to a central computer operations centre).

Until the early 1980s all leaselines were *analogue* leaselines, rented according to capacity or *bandwidth*. Today nearly all network equipment is digital equipment, and *digital leaselines* are more common. These are based upon *TDM (time division multiplex)* technology so that a digital leaseline is realized as a TDM *channel*. The capacity of a TDM *channel* is quoted in terms of its *bitrate*.

In the days of analogue leaselines, the leaselines were in effect *private wires* laid between the two desired endpoints and dedicated to a single, customer's use. Sometimes indeed the realization was a cable dedicated to the customer's use. In other cases the physical cables were shared between a number of customers and carried several channels. *Multiplexors* and *crossconnects* provided for the shared use of the physical lines but also ensured the separation and privacy of bandwidth allocated to individual customers.

Digital leaselines rely on transmission networks built from TDM *multiplexors* and *crossconnects*. These provide for the sharing of capacity on digital transmission lines and for the connection of individually dedicated *channels* of appropriate *bitrates* between two desired endpoints. The *multiplexors* provide for the sharing of lines by combining several *channels* together for carriage across the same physical line by interleaving the bits corresponding to the separate channels. The *crossconnects* provide for the interconnection of channels on separate line systems without requiring the intermediate demultiplexing of the individual channels (Figure 4.1).

The *multiplexor* and *crossconnect* terminology has also made its way into the ATM vocabulary, and ATM multiplexors and crossconnects play a similar role to their TDM relatives in combining and crossconnecting ATM connections. By means of these devices and the ATM *constant bit rate (CBR)* or *circuit emulation (CES)* service, ATM can provide a service nearby equivalent to TDM leaselines. We next assess the advantages and disadvantages of replacing a digital leaseline with the ATM CBR service.

The three main historical justifications for using leaselines are:

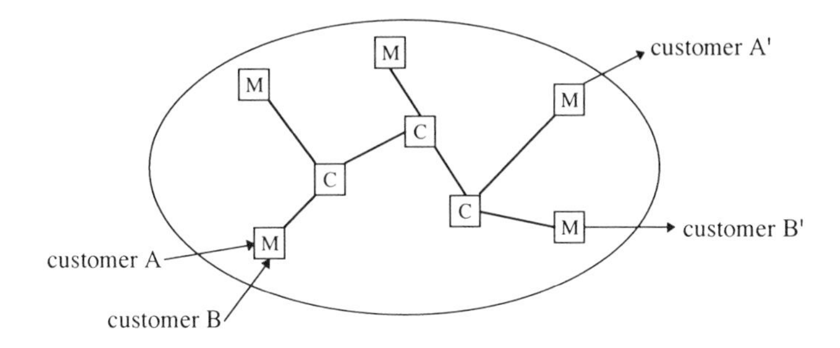

Figure 4.1 Function of multiplexors (M) and crossconnects (C)

- financial economy;
- technical performance requirements of the end devices;
- security (i.e. privacy) reasons.

Financial economy (leaseline versus ATM CBR)

Where a company generates daily more than about 70 minutes of telephone calls between two particular sites, it typically makes economic sense to have a *leaseline* installed between these two sites. The flat charge leaseline rental has to be paid, but there are no further usage-dependent telephone charges for calls made.

However, where all 70 minutes per day is generated by seven people talking for the same ten minutes (e.g. 09.30 till 09.40 each day) then seven separate leaselines would be needed, and using the public telephone service may remain more economic than the leaselines.

A CBR service offered by a public ATM network carrier is also likely to bear a flat charge, but be considerably cheaper than the equivalent leaseline. There may, however, be additional volume-dependent charges to be paid. The ATM CBR service is thus likely to be more economic than a leaseline where lower volumes of information are to be carried, but the leaseline may continue to pay in at high volumes. Returning to our example, the leaseline may continue to be the cheapest option where a single leaseline is able to carry the 70 minutes of traffic spread throughout the day. But where the 70 minutes are generated by the seven simultaneous 10-minute calls, it may be cheaper to rent seven ATM CBR channels from a public ATM network carrier than to pay the telephone charges or the costs of seven leaselines.

Alternatively, a corporate network manager might like to consider equipping an existing, but not fully utilized, point-to-point leased line with an ATM or similar cell relay switching device (e.g. Stratacom) to provide for simultaneous line usage by other devices. He can then make more use of the line. Figure 4.2 shows a possible arrangement. Existing end devices are connected via the ATM switches by means of the ATM CBR service or equivalent. New devices are also connected. The existing leaseline becomes the trunk between the ATM switches.

The statistical multiplexing of lines as illustrated by Figure 4.2 is only feasible and economic where the statistical sum of the line loadings generated by each of the end devices is less than the total carrying capacity (i.e. bitrate) of the leaseline.

Technical performance (leaseline versus ATM CBR)

In the pre-ATM world, leaselines were sometimes unavoidable where a very high quality and reliability of line was needed between two communicating devices situated in different sites. Some highly sensitive and time-critical applications have historically demanded leaselines because of the problems of assuring throughput rates through, for example, packet switching and frame relay networks. For some applications the jitter rate (i.e. the variability in speed of arrival of information bits at the destination) of packet switching and frame relay networks is unacceptable. For many applications the telephone network did not provide an alternative because of the limited bitrates possible

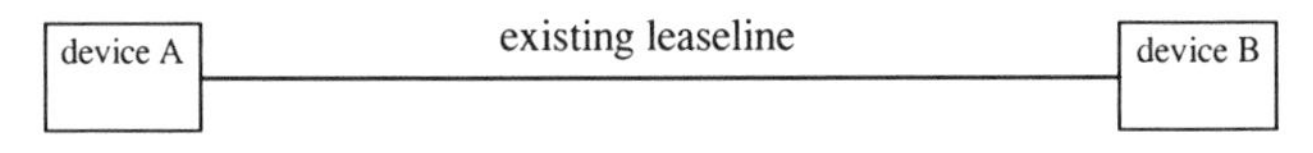

(a) Existing direct connection of devices

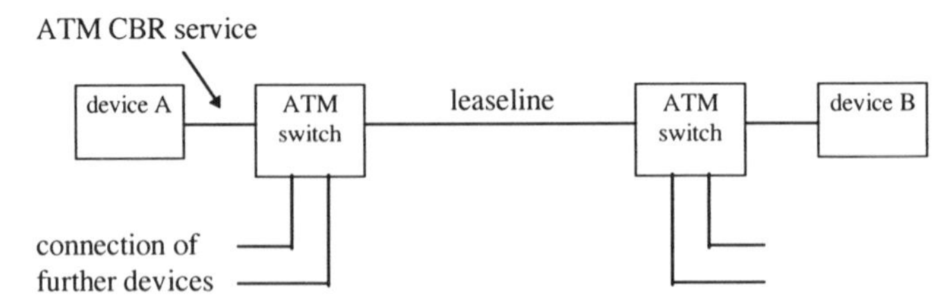

(b) Improved unilization of the line by ATM statistical multiplexing of further devices

Figure 4.2 Use of the ATM CBR service for improving line utilization by statistical multiplexing

with dial-up modems, and the problem with ISDN was the bit error rates caused by the digital telephone switches which make up an ISDN.

For some of those applications which previously demanded a leaseline the ATM CBR service will provide a cheaper and viable technical alternative (because of the high speed of ATM). But for the most critical of applications a clear channel leaseline may remain the only viable option. This is because while ATM is specifically designed to offer very low jitter on CBR service, the statistical nature of the cell relay technique behind ATM means that jitter cannot be eliminated. Users are therefore best advised to test their most critical applications over ATM CBR service before full commitment or investment.

Security (leaseline versus ATM CBR)

A common network planning philosophy of some corporate network managers is to use leaselines in preference to public switched networks for cases where confidential information needs to be conveyed between two particular sites.

The use of direct leaselines (as opposed to public data networks) for connecting private Local Area Networks (LANs) connected using routers and the TCP/IP (i.e. *transmission control protocol/internet protocol*) provides protection against break-in by the *hackers* of the *Internet* world. However, to believe that a leaseline (or *private wire*) provides protection against physical 'tapping' of a line is a misguided dream. In many ways a private wire is easier to tap than a public switched network. To do so you need only find the appropriate physical line in the right manhole, and then 'tap' into it.

Switched networks (including ATM) can be made relatively hard to 'tap', by avoiding a hard and fast route through the network being used – the connection is occasionally broken and then re-established.

4.3 TDM Networks versus ATM Constant Bit Rate (CBR) Service

TDM networks are generally set up by companies for one of two purposes:

- for in-house or campus cabling of large sites; or
- for the establishment of private leaseline-like networks interconnecting company sites spread over a wide geographic area.

The use of TDM technology in in-house or on-campus networks is generally to get around the problems associated with direct cabling as explained earlier. A typical usage of such TDM equipment might be for the multiplexing of a number of copper pair cables of a building's *structured wiring system* onto the fibre cables which may run from technical rooms in the basement up through the cabling *risers*. It is common, for example, for computer and telephone switch equipment to be housed in building basements, from where individual channels are multiplexed together to run over fibre cables up the building risers to wiring cabinets on each storey. In these wiring cabinets are TDM equipments which demultiplex the individual channels onto copper cable pairs which run out to telephone and computer sockets in individual offices (Figure 4.3).

The replacement of TDM multiplexors with ATM devices for in-house cabling usage is likely to be one of the first application fields of ATM. ATM will bring much more flexibility in the range of bandwidths available for in-house usage, and will particularly benefit high bandwidth data applications, like LAN–LAN interconnections or *client/server* (e.g. *UNIX*) computer networks. This in turn is likely to pressure the development of ATM PBXs.

In the *wide area* world, TDM has historically provided a means for cost savings on leaseline networks, by enabling companies to buy relatively high bandwidth leaselines at wholesale prices and then break them down into individual user channels at lower bit rates. ATM multiplexors could provide a similar functionality. The advantage of using ATM rather than TDM is the extra efficiency gained through the statistical multiplexing technique. The efficiency gain is most marked where the line carries mostly data communication.

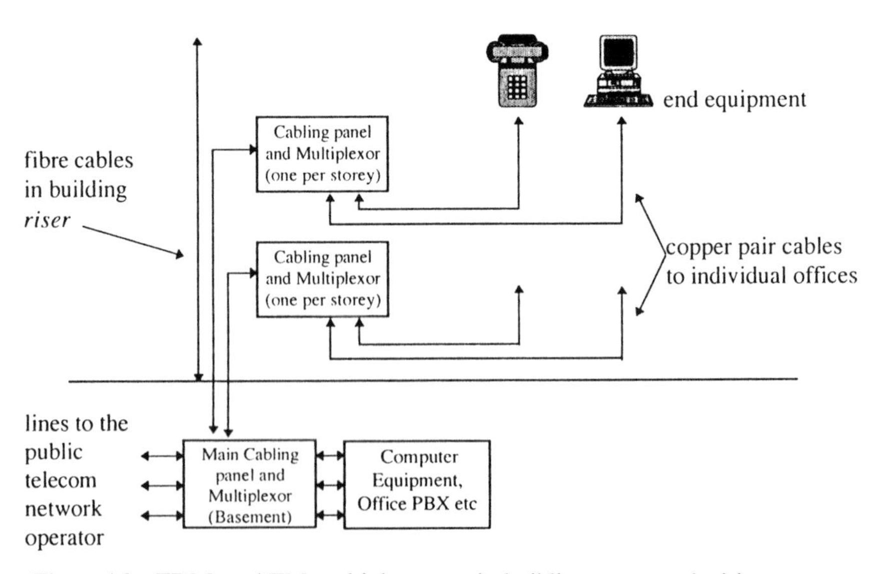

Figure 4.3 TDM or ATM multiplexor use in building structured wiring systems

The drawback of ATM is that the standards do not define trunk interfaces below 2 Mbit/s. Thus where the existing wide area line is of a rate lower than 2 Mbit/s it may be wise to stick with TDM. Alternatively, another solution may be to use 'ATM'-like multiplexors which use a 'proprietary' (i.e. manufacturer specific) method to offer bitrates lower than 2 Mbit/s. Such devices are Northern Telecom's (Nortel's) *Magellan Passport*, IBM's *Nways(2220)*, and Stratacom's *IPX*, *IGX* or *BPX*.

4.4 X.25 Packet Switching versus ATM

X.25 packet switching was the first universal system for connecting all types of data communications devices onto a common, switched data network. It was developed in the 1970s, when the relative speed of the communicating devices was very low (in comparison with today's devices) and the quality of wide area digital lines was comparatively poor. As a result (and to their credit) X.25 packet networks are highly robust against poor line quality. In short, X.25 networks are able to survive and even recover from even extensive bit errors on digital lines. The problem is that the cost of this robustness is the very limited line speeds which are possible, and the relative inefficiency of line utilization in the case of higher quality lines.

The problems which arise when attempting to operate X.25 protocol at high speeds are due to the *windowing* technique employed by X.25 to help avoid errors. To illustrate the problem we consider trying to use a 2 Mbit/s line to carry X.25 data over a distance of 1000 km.

As Figure 4.4 illustrates, on a high speed data transmission line, there are always a large number of bits in transit on the line at any point in time (because of its length) – in our example around 20000 bits or 2500 bytes (line length × bitrate/speed of light). That should blow any preconception you might have had that electricity travels so fast that we can consider sender and transmitter to be in synchronism with one another! These bits in transit on the line must be considered when designing high speed data networks, if the network is to operate efficiently.

X.25 lays a very high priority upon the safe arrival of bits, in the correct order and without errors. One of the methods used to ensure safe arrival is the use of an acknowledgement *window*. Only so many packets (as defined by the *window size* – typically up to 7) may be transmitted by the sending device before an acknowledgement is received confirming safe arrival. Since the typical maximum packet size is defined as 256 bytes, this means only around 1800 bits (7 × 256) may be transmitted by the sender before an acknowledgement is returned by the receiver to confirm safe arrival. Even before considering the inefficiencies caused by packet *overheads*, the X.25 *window* will constrain the

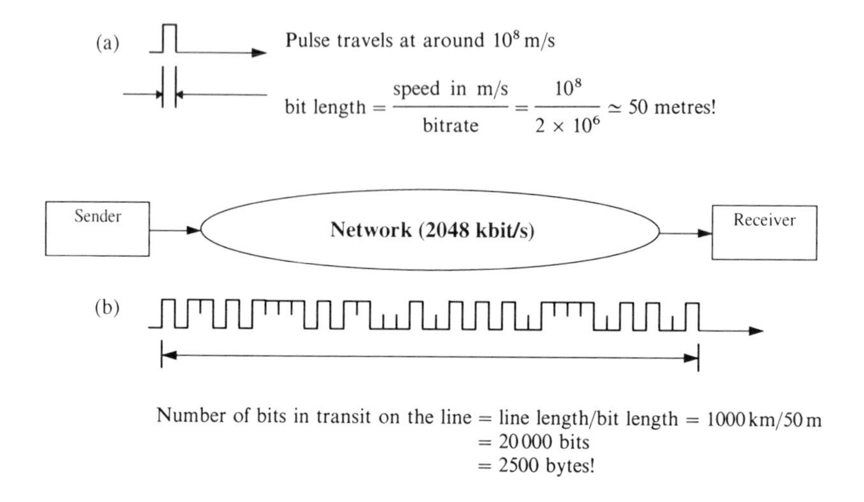

Figure 4.4 Bits in transit in a high-speed data network

efficiency of the line of Figure 4.4 to a maximum of 1800/2500 (maximum bits allowed in transit/available bits in transit) – or around 70 per cent.

An immediate reaction to the throughput problem caused by the restricted size of the X.25 *window* (typically 7 packets) is to increase the maximum window size. Unfortunately this only generates new problems. First, the end devices need to provide much greater storage buffers for retaining copies of the sent but unacknowledged information. Second, because the window size is greater, so is the likelihood of errors within a window. The probability of the need for a retransmission of the information in order to eliminate the errors is thus also greater. Also, because of the increased window size, the time required for retransmission is longer. The conclusion is that increasing the window size does not necessarily increase throughput, since the capacity can become tied up with retransmission of data. Indeed the throughput may reduce.

Today's digital transmission is by several orders of magnitude of better quality than that of the 1970s, so that the heavy duty error detection and correction techniques used by X.25 have become redundant. Modern data communications *protocols* (e.g. *frame relay, ATM*) tend not to use such heavy duty error detection and correction techniques and therefore are generally more efficient in terms of possible information throughput than X.25 at higher bitrates.

X.25 remains important because of its ubiquitous potential for interconnecting computer and other devices of many different manufacturers and types and because of its strong market presence. It is, however, no longer favoured for applications or devices needing bitrates much above 64 kbit/s. At 64 kbit/s *frame relay* is today's preferred method. For user rates above 2 Mbit/s, native

ATM (UNI) should be considered (provided, of course, that the end device supports it, or some form of ATM converting device is used).

4.5 Frame Relay versus ATM

The *frame relay* protocol was developed as a result of the performance difficulties of X.25, and is based heavily on X.25. In effect *frame relay* is a simpler form of X.25 in which most of the error detection and correction responsibilities have been removed from the intermediate network devices. Instead the end devices communicating with one another are obliged to take greater responsibility for checking the error-free nature of received information.

Higher quality modern networks have made the robustness of X.25 largely redundant. By reducing the robustness to error, the packet (or correctly *frame*) overhead has been reduced, adding dramatically to the efficiency and throughput of the circuit. Meanwhile, the similarity of the techniques has enabled manufacturers of X.25 network switch equipment to undertake only marginal developments in order to adapt their devices for support of the *frame relay* protocol.

Frame relay has established itself as the standard method for connecting data communications devices which require interconnecting links of bitrates ranging between 64 kbit/s and about 2 Mbit/s. It is thus already the standard method for permanently interconnecting LANs (*local area networks*) in different buildings or sites, where it provides for a high speed link between LAN *routers* (e.g. *Cisco* or *Wellfleet*).

Unlike X.25 packet switching, which it originally appeared to shun in favour of its own *SNA (systems network architecture)* protocol and never wholeheartedly supported, the IBM company has been one of the early adopters of frame relay. Frame relay will be one of the standard higher speed communications interfaces for connecting to IBM mainframe computers.

The main drawbacks to date of frame relay have been:

- the inability of frame relay networks to switch connections, instead being only capable of supporting point-to-point connections (*PVC*s, or *permanent virtual circuits*);
- its inability to support very high speed connections, particularly those demanding low jitter performance (e.g. those involving video or moving images).

The limitation to *PVC* connections has meant that frame relay networks have had to bear rather rigid topologies, without much flexibility for direct connection to a range of distant devices as is possible in a switched network.

Recently (in 1994) the standards for *SVC (switched virtual circuit)* frame relay have been agreed and equipment with the new capability is appearing. This will make it possible for frame relay devices to 'dial' different remote devices rather than only always communicating with a single partner or small number of partners. As we illustrate in Figure 4.5 this will lead to somewhat simpler network topologies and reduced port hardware needs, and thus to easier network planning.

In the example of Figure 4.5(a), four LANs are interconnected using routers and a full mesh topology of six PVC frame relay connections. This requires only one physical port at each of the routers but the configuration of three separate *logical channels* to be carried by this port. Where a very large number of PVCs are needed for the full mesh, i.e. more than 1024, additional hardware ports may also be needed.

In contrast, in Figure 4.5(b) the same devices are interconnected using SVC frame relay connections. In this case, any of the routers can establish switched connections on demand to each of the other routers. As far as the routers are concerned they still appear to be fully meshed with one another, but now the job of administering the network topology has become much simpler (far fewer permanent connections to maintain and to reconfigure when the network topology changes). Furthermore, should any of the routers fail, then a new path to a back-up device can be established.

Because of its efficiency in supporting data applications, frame relay protocol is becoming a well-established networking technique in corporate networks and a wide range of reliable equipment is available. This strong position will be reinforced by the introduction of SVC service, which will increase the flexibility of frame relay networks. Therefore in existing data networks with wide area links not requiring more than 2 Mbit/s capacity on any single connection, frame relay is unlikely to be substituted in the short term for ATM.

ATM is poorly equipped to compete with frame relay at rates under 2 Mbit/s because the ATM cell size of 53 bytes is too inflexible to be effective at

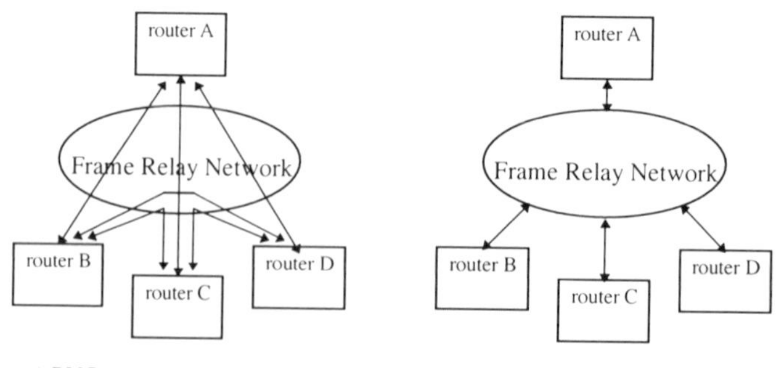

Figure 4.5 PVC and SVC frame relay services compared

such low bitrates. The cell duration, for example, at 512 kbit/s is about 1 ms (compared with 12 μs at 34 Mbit/s), so that there is a much greater risk of unacceptable delay of cells waiting to be sent. Jitter (the variability of the delay) is also likely to be a problem for jitter-sensitive applications like voice. Meanwhile, even for data applications, the extra overhead of the cell format is adding to overall delay and impairing network throughput.

ATM will displace frame relay in new data networks requiring very high bandwidths (from 2 Mbit/s), particularly where these networks are required to support applications with low jitter needs (e.g. moving image or video). Also driving the demand for much higher bandwidth connections are new 'networked' applications requiring the high speed interconnection of LANs or of *client* devices with computer *servers*. This is causing turmoil in the datacommuncations and router industry where the manufacturers are all desperate to be amongst the first to introduce ATM technology. The mere availability of ATM compatible end devices will create demand for ATM network services and cause substitution of frame relay.

Recognizing the complementary strengths of frame relay for lower speed data and ATM for higher bitrates, the standards incorporate ATM/frame relay interworking and some manufacturers are developing network switching devices which provide for mixed ATM and frame relay. Such devices (e.g. Northern Telecom's (Nortel's) *Magellan Passport* and IBM's *Nways*) may thus become common in corporate data networks as companies make the transition to ATM.

4.6 LANs and FDDI versus ATM

Token ring and *ethernet* LANs are today's standard method of networking personal computers (PCs), printers, data storage servers and other devices in office environments. They allow individual employees to share a common pool of computing resources. The early LANs which appeared in the mid-1980s operated at bitspeeds up to 10 Mbit/s. At the time this was considered a very high bitrate. In the meantime the processing speed capabilities of the PCs connected to such LANs have increased dramatically, and the prevalence of such networks has led to a boom in networked applications, where large volumes of data are continuously being communicated across the network. In consequence LAN speeds have been increased (up to 16 Mbit/s) and very high speed network technologies (around 100 Mbit/s) have emerged for interconnecting multiple LANs within large building or campus sites via fibre cabling. One such technology is *FDDI (fibre distributed data interface)*. A typical LAN campus network is illustrated in Figure 4.6.

FDDI has been a mild success for high-speed interconnection of LAN

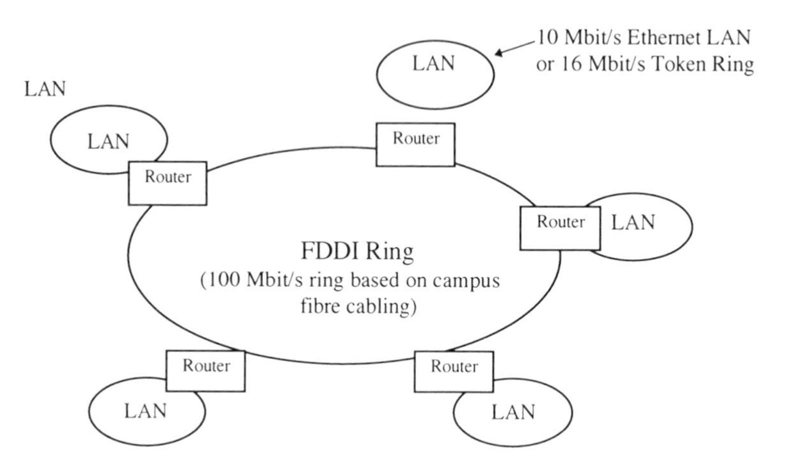

Figure 4.6 Typical campus network of meshed LANs

routers on campus sites, but has never successfully made the transition to become a more general purpose telecommunications medium. There have been attempts to adapt it to carry internal office telephone lines (FDDI-2) but these were largely unsuccessful. Its limited scope has led to its early obsolescence and few, if any, manufacturers who do not already offer FDDI-compatible equipment are likely to add it to their portfolio.

ATM stands in direct competition to FDDI's one-market niche but can also offer much more. First, it offers the choice of either 34 Mbit/s or 155 Mbit/s bitrates; second, it is a highly effective means of data communication; third, its much wider scope of use (e.g. for voice, video, etc.) affords it much greater future potential than FDDI. The consequence is that the earliest available ATM devices have aggressively attacked the campus LAN interconnect market.

The next inevitable assault will be on the replacement of the Ethernet and Token Ring fabric used within the individual LANs. This has already commenced, with the publication by the ATM Forum of standards for ATM *LAN emulation (LANE)*. ATM *LANE* networks are intended to replace traditional ethernet and token ring alternatives.

4.7 SMDS (DQDB) versus ATM

SMDS (*switched multimegabit digital service* – in Germany *Datex-M*) is a network service based on the DQDB (*distributed queue dual bus*) technique defined by IEEE standard 802.6. SMDS has been available as a public service in some countries (particularly North America) since 1991. The DQDB protocol was developed by Telecom Australia, the University of Western Australia and their jointly owned company, QPSX Communications Limited. DQDB offers a reliable data switching service, particularly well suited for LAN interconnection and similar applications which require metropolitan interconnection of different buildings or sites, hence the appearance of the terminology *MAN (metropolitan area network)*.

DQDB was originally foreseen as the next step towards broadband ISDN. Unfortunately, in its early days, users were unsure of its data networking benefits relative to those of FDDI. More lately its capabilities for broadband and multiservice switching have been overshadowed by ATM, so that it seems deemed to obsolescence even before maturity. Few global manufacturers and public network operators (especially those outside North America) who have not already committed themselves to DQDB developments are likely to do so.

Even though the SMDS standards and protocols (e.g. *SIP, subscriber interface protocol*; *DXI, data eXchange interface*) have been defined in ATM standards to interwork and run via means of ATM transport networks, it is more likely to be economic to re-equip end devices to communicate in native ATM protocol rather than adapt the ATM network for SMDS (unless absolutely necessary).

4.8 Telephone Networks versus ATM

To date the main telecommunications investment of most public network carriers and indeed many private corporate network operators has been centred on telephone services. Public operators have centred their activities around the *public switched telephone network (PSTN)*, and corporations have invested in office telephone systems, including private branch exchanges (PBX) and fax machines. Undoubtedly there have been sound reasons for such investment, and indeed telephone and fax will remain amongst the most important business communication tools. The technology used to support telephone network services, however, will move away from classical *circuit switching* techniques to more efficient and flexible techniques, including, foremost, ATM.

The method by which speech, fax and other telephone-type connections should be carried and switched by ATM networks is not yet fully standardized, but what is already clear is that the cell relay technique behind ATM offers two irresistible attributes for the line usage and cost optimization of telephone networks:

- voice compression
- silence suppression

As shown in Figure 4.7, voice compression contributes to the efficiency of a network by reducing the bitrate required to be carried by the network – particularly across the costly long distance lines. The particular advantage of ATM networks compared to traditional telephone networks is that they can also switch the connections at the compressed bitrate. In traditional telephone networks, as we discovered in chapter 3, the signals must be decompressed for each switching stage. This adds to the cost and complexity of the network, and has a degrading effect on the end-to-end connection quality. Although ATM networks inflict voice connections with more jitter than traditional telephone networks, this may be more than compensated by the improved quality resulting from fewer decompression/recompressions at the various switching points.

Silence suppression is (Figure 4.8) a natural by-product of the cell relay technique used by ATM. The signals representing bursts of speech (say individual words) are packed into cells and carried across the network. But

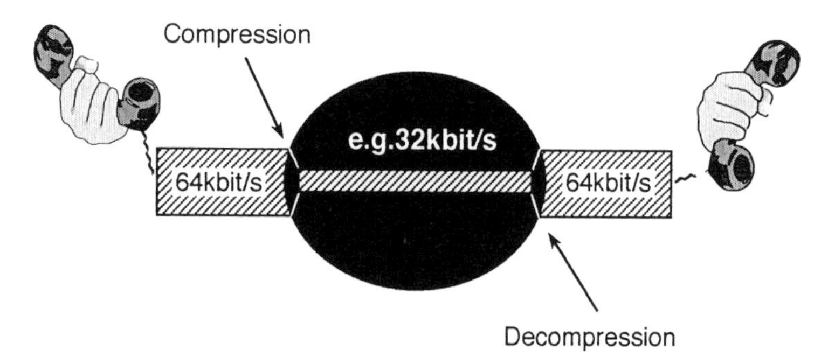

Figure 4.7 Network bandwidth saving by means of voice compression

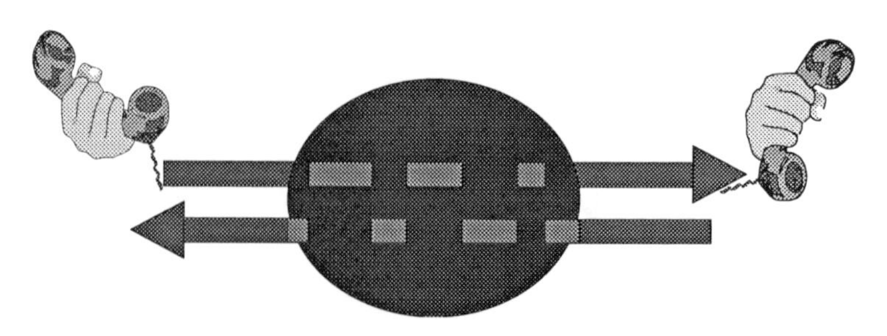

Figure 4.8 Network bandwidth saving by means of silence suppression

when there is silence there is no need to carry a signal across the network, so cells can be saved. This is the principle of silence suppression. These cells can then be used for other conversations or data connections.

Using silence suppression, only slightly more network capacity is needed than the total of the average bitrates of each of the individual connections. Thus a capacity saving is achieved in the network. Because in normal conversation only one person talks at a time a bandwidth saving of at least 50 per cent is possible (since one or other of the transmit or receive paths is always idle). In practice the saving is around 70 per cent, since there are also silence periods between words. Combined with 2:1 (i.e. 32 kbit/s) voice compression, the saving could be as high as 85 per cent of the bitrate!

In traditional telephone networks silence suppression is not possible because the connection is switched permanently through for the entire duration of the call.

4.9 ISDN versus ATM

ISDN (integrated services digital network) is a technology based upon the telephone network, but capable of offering *integrated* services. An *integrated service network* allows different types of signals and information (e.g. telephone, facsimile, PC, mainframe computer, video) to be carried simultaneously over the same connection line. This type of network allows a company to optimize its telecommunication network usage by sharing the same lines for different purposes. During the daytime the network can be used largely for telephone conversations and PC or LAN networking, while overnight the same lines may be reused for computer or other applications (e.g. consolidation of daily retail sales, broadcast of new price lists to stores, despatch of delivery notices to warehouses, transfer of manufacturing specifications, video film transfer). Alternatively, by dialling-up connections across an integrated service network only for the periods needed, a customer could minimize his or her network usage and the related costs.

There are two technologies capable of providing the basis for wide area integrated service networks. These are ISDN and ATM.

ISDN has been designed to enable computer networking and relatively simple video transmission to be conducted across today's digital telephone networks. In order to achieve this, ISDN telephones must be able to signal to the network not only the desired destination of the call (the dialled telephone number), but also the type of connection which is required. The traditional pulsing of individual dialled digits from telephone to network no longer provides sufficient information for the call setup and a much more advanced ISDN signalling system has evolved.

Apart from providing for the dial-up demand of either telephone, computer or videotelephone connections, ISDN signalling has a number of additional spin-off (correctly called *supplementary service*) benefits. These include the capability to identify a telephone caller's number prior to answer (this could be useful to direct specific customers to specific customer service agents or to dissuade malicious callers), the capability to 'ring back when free' and the capability to conduct conference calls involving three or more callers. In addition, handsets supporting the latest European version of ISDN *user-to-network signalling* (Euro-ISDN or EDSS-1) will be capable of operating in any European country.

ISDN, however, is limited by its telephone network ancestry. The problem with traditional telephone networks and ISDN is that all network connections have the same carrying capacity (64 kbit/s) – that needed for a telephone call. For some applications (e.g. monitoring building alarms) this is far more capacity than is needed, while for some of the latest PC and LAN interconnection methods and for video applications, this capacity is far too small. That is not to say that the overall capacity does not exist within an ISDN, but rather like a potato going through a chip machine, the signal would first have to be broken up for transmission and then reassembled as a single potato (with all the matching complications) at the destination.

ATM does not suffer these difficulties. Rather than being a fixed capacity switching technique, ATM can switch connections of almost any bitrate. Thus ATM can support simultaneously telephone, computer or LAN connections and video. The best analogy to an ATM network is a highway road network. The highways are analogous to ATM transmission lines – very high capacity connections. The highway interchanges are the ATM switches. The traffic on the road represents the different types of telecommunication traffic. In an ISDN network, only cars of a single specific type are able to travel. If this size is too big, then the vehicle travels half empty, otherwise the load must be split between several vehicles. On the other hand, in an ATM network a range of different sized vehicles are available. It remains only for the customer to decide when starting his journey (i.e. when setting up his connection) whether he needs a mini or a juggernaut. Because the customer may select his exact capacity needs prior to each call, the efficiency of the network is maximized and the congestion kept to a minimum (no trucks returning empty or running only half full). Perhaps ATM is what the politicians have in mind when they speak of the information highway!

The supplementary services supported by ISDN, e.g. *calling line identity* – (the delivery of the caller's number even before answer) or *ring back when free* (later connection to a currently busy line), will also be supported by ATM, because the network access signalling, Q.2931 at the ATM UNI (*user network interface*), is based on the ISDN user network signalling (ITU-T recommendation Q.931).

4.10 Intelligent Network (IN) versus ATM

The *intelligent network (IN)* is a development resulting from the difficulties experienced in managing complex services within extensive telephone networks. In the mid-1980s, fuelled by the desire to establish differentiation from their competitors, public network operators began to offer a range of new, more complicated services including, amongst others, 800 service (freephone) and *VPN (virtual private network)*. 800 service offered for the ability for the called party (rather than the caller) to pay for the call. VPN enabled companies to operate company-specific telephone numbering schemes (as is possible in a private telephone network) while simultaneously benefitting from the economic benefits of large scale public telephone networks (in short a *virtual* or 'apparent' private network, but embedded within the public telephone network).

Initially 800 service and VPN service were realized using software and configuration data spread throughout all the telephone switches in the operator's network. But it quickly became apparent that this was unmanageable. The effort required to keep all the various data tables in each of the individual switches both up-to-date and consistent with each of the other switches was simply too great. The solution was IN.

The clever feature of IN *(intelligent network)* is the centralization of the most complex service software and configuration data into a single central database, where it can be more easily managed. When individual telephone switches encounter calls requiring intelligent network services (such as 800 or VPN) then the call setup is interrupted and an enquiry is made to the central database (the *SCP* or *service control point*) to determine how the call setup should be dealt with. The SCP replies to the *SSP (service switching point* – the telephone switch making the SCP enquiry) with instructions on how to handle and bill the call (Figure 4.9).

In an IN realization of 800 service the caller dials the 800 (freephone) number, e.g. 800 425 XXXX, which is first recognized by the SSP to be an intelligent network service. Call setup is temporarily interrupted by the SSP while an enquiry to the SCP is *triggered*. In response to the enquiry, the SCP returns instructions to the SSP to forward the call to a real network address (e.g. 212 123 XXXX – a telephone line in New York, USA) and to charge the call to this receiving number.

IN realization prevents, in the case of our example above, every single telephone exchange in the network needing to know that number 800 425 XXXX should be translated and routed as if the number 212 123 XXXX had been dialled and then billed to the called party. Instead this information must be stored only once – at the SCP. In addition, *intelligent network* technology will make a range of new types of services available and drastically reduce the development time needed for preparing new services for general release (since

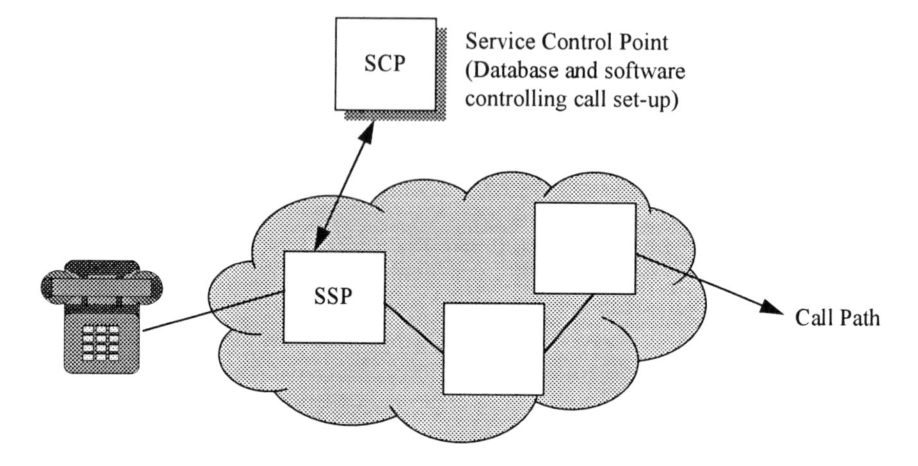

Figure 4.9 Elements of the Intelligent Network

these may only need to be developed for the SCP). Examples of services made possible by *intelligent networks* include freephone service (service 800), information services (services 700 and 900), calling card service and 'find-me'.

Freephone service is already enabling companies to reduce the barriers which prevent customers contacting them, by paying for telephone calls as the receiving party. Freephone number usage is already common in many service and sales departments.

Information service, in contrast to freephone, allows the call recipient to charge a premium for the call, allowing the recipient to invoice the caller not only for the call itself but also for information which he received during the call. Thus a hotline service, for example, might give the latest weather forecast or sports results.

Calling card service may be of interest to companies with travelling employees who wish to avoid the surcharges of hotels or the inconvenience of public payphones. The calling card allows the employee to make his calls from any convenient phone while maintaining a strict control of costs through a monthly itemized bill. 'Find-me' service is intended to simplify the calling of hard-to-reach employees, trying in turn perhaps different alternatives (e.g. telephone, mobile phone, pager, voicemail).

Currently the realization of intelligent networks is largely limited to telephone networks, but the principle of enabling more sophisticated *service scripts* and centralized configuration data (e.g. number translation tables) is also attractive in improving the functionality and manageability of other types of networks. A combination of IN and ATM would indeed be a powerful combination – and most of the world's major network equipment manufacturers are determined to be the first to be able to offer it.

On top of the highly flexible and efficient switching fabric offered by ATM, *intelligent network* technology has the capability to greatly enhance the control and management of the network in real-time. A *virtual private network (VPN)* created from ATM and IN would offer enormous data, voice and bandwidth switching range with the security and individuality of a private company telecommunications network (including bespoke company-internal telephone numbering scheme) but with the additional economic benefits of resource sharing which implementation on a public network can offer. Thus a corporation used to operating a private network for its flexibility and cost-saving benefits can achieve its goal without the need for heavy investment in technology or operating infrastructure.

In combination ATM and IN will help bring about a dramatic reduction in lead times for new connections and new services, will afford the customer a stronger role in network management without the burden of network operation, and will enable more robust load-balancing and self-healing networks to become reality.

5

ATM's Technical Specifications

Much effort is being invested by the world's network operators and equipment manufacturers in developing the standards and specifications for ATM. The standards are emerging in a range of different standards bodies and are still far from complete, but it is nonetheless important for all parties contemplating developing or using ATM to understand them. This chapter provides first a review of the institutions involved in developing standards, and reviews in particular the standards of greatest worldwide significance – those of ATM Forum and of ITU-T (International Telecommunications Union – Telecommunications Standardization Sector).

5.1 Standards Bodies

There are four bodies prominent in the development of ATM standards. These are, first and foremost, the *ATM Forum* and the world-recognized United Nations agency body, *ITU-T*. In addition, the regional standards bodies, e.g. *ANSI* and *ETSI*, are also playing an important role, particularly in identifying and resolving problems during implementation.

The ATM Forum came about through the interest of four major manufacturers who wanted to speed up the standardization process of ATM. It was set up in October 1991 by Northern Telecom (now called Nortel), Sprint, Sun Microsystems and Digital Equipment Corporation (DEC). In January 1992 membership was opened to wider participation from the telecommunication industry. The purpose of the ATM Forum is to promote the development of technical standards for ATM and to raise awareness in the market of its capabilities. Many ATM standards adopted by ITU-T are heavily based on (if not identical with) with those previously issued by the ATM Forum, but have tended to be published up to two years later. This can lead to confusion.

Recently, a slight divergence between ATM Forum and ITU-T has occurred. The ATM Forum appears to be more oriented to equipment manufacturers and *Internet*, thus tending to concentrate most on private ATM networks, while ITU-T retains its public network operator focus and OSI (Open Systems Interconnection) orientation. As a result, it is likely that ATM Forum will promote short term, pragmatic solutions, while ITU-T is likely to generate robust standards, suitable for public networks.

As an example the ATM Forum has chosen *SNMP (simple network management protocol* – a heritage of the *Internet* world) as the basis for the *ILMI (interim local management interface)*. The longer-term solution standardized by ITU-T will almost certainly be based on *CMIP (common management information protocol)*. CMIP is the OSI equivalent of SNMP. The ATM forum has thus suggested a pragmatic solution to start with, while those waiting for the full ITU-T solution will have to wait a little longer.

ANSI (American National Standards Institute) and *ETSI (European Telecommunications Standards Institute)* are regional standards-setting bodies which are active in the development of telecommunications standards. They are significant because of their strong influence (through both market force and regulatory edict) on the markets in North America and the European Community. However, neither ANSI nor ETSI is primarily focused on ATM standards, so they too are tending to base their work on output from the ATM Forum.

5.2 The ATM Forum

The work of the ATM Forum is split into three main areas: Technical, Marketing and User need analysis, and the working committees of the forum are split into these areas. Technical committee meetings are those in which the technical standards are developed by the *principal members* of the forum. The *market awareness and education (MA&E)* committee then takes over the responsibility of educational material, tutorials, press releases and presentations needed to improve the general telecommunications market awareness of the capabilities of ATM. These meetings, too, may only be attended by *principal members. Auditing members* are not allowed to attend meetings of the *forum* but receive copies of the published documents. Finally, the *end user roundtable (ENR)* was set up in August 1993 for a new class of members, *users*. This committee is reserved only for *user members*, and is dedicated to the task of developing better understanding of users' basic needs of ATM.

The ATM Forum may be contacted at the following address:

The ATM Forum
2570 West El Camino Real
Suite 304
Mountain View
California CA 94040
United States of America
Telephone: +(1) 415 949 6700
Internet: info@atmforum.com

5.3 ITU-T (*International Telecommunications Union – Telecommunications Standardization Sector*)

ITU-T, or *CCITT, International Telephone and Telegraph Consultative Committee*, as it used to be known, is the world's leading telecommunications standards authority. Technical standards for ATM appear primarily amongst its I- and Q-series *recommendations* for *broadband-ISDN (B-ISDN)*. ITU-T may be contacted at the following address:

International Telecommunications Union
Place des Nations
CH-1211
Génève 20
Switzerland
Telephone: +(41) 22 730 5111

Alternatively, ITU standards may be viewed directly over Internet by first contacting ITU helpdesk at:
sales@itu.ch

5.4 ANSI (American National Standards Institute)

ANSI, and in particular the *T1 committee*, provide for North American debate of telecommunications network standards prior to worldwide agreement at ITU level. Subsequently ANSI adapts ITU-T standards to the particular market environments of the USA and the particular transmission equipment standards (SONET as opposed to SDH) which prevail in North America. ANSI may be contacted at the following address:

American National Standards Institute
1430 Broadway
New York
NY 10018
United States of America
Telephone: +(1) 212 354 3300

T1 Committee
Exchange Carriers Standards Association
5430 Grosvenor Lane
3200 Bethesda
Maryland MD20814–2122
United States of America
Telephone: +(1) 301 564 4504

5.5 ETSI (European Telecommunications Standards Institute)

ETSI is the most important of European bodies producing technical standards for the telecommunications industry. Its importance lies in its wide support from European governments and in the mandatory use of many of its standards as decreed by Parliament and the Commission of the European Union.
ETSI standards appear either *as ETSs (European telecommunications standards)* or as *NETs (normes européennes de télécommunications)*. ETSs are standards defining the funtionality, interfaces and protocols of telecommunications networks. NETs are mandatory standards adopted by the European Union as the basis for network terminal equipment approval. ETSI also publishes technical guidelines for network design and operations. These appear as *ETRs (ETSI technical reports)*. Finally, in its role serving the European regulation of telecommunications, a number of documents called *TBRs (technical bases for regulations)* are also published. ETSI may be contacted at the following address.

European Telecommunications Standards Institute
F06921 Sophia-Antipolis Cedex
France
Telephone:+(33) 92 94 4200
Facsimile:+(33) 93 65 4716

5.6 Overview of the Standards

There are numerous standards for ATM. Unfortunately, however, it is not easy for a newcomer or an outsider to the subject to understand them when reading them in isolation. The structure of the standards is confused because of the various issuing bodies. To make matters worse, even the sets of standards issued by single bodies are not always clearly structured or numbered.

Because of the continuing rapid evolution of the standards, more recently-issued documents may to some extent contradict earlier documents – as a result of further research and definition work. Also problematic for the reader can be the extensive use of acronyms – particularly when different terms are used by different standards bodies to mean the same thing (e.g. *customer equipment* versus *user device*).

It is, quite simply, difficult to determine which specification to read first, and how the various different specifications relate to one another. For this reason, appendix 2 to this book presents a full listing of published ITU-T and ATM forum standards and is followed by a comprehensive glossary of acronyms as current at December 1995. Recommendations and standards are referenced by their recommendation number and title (e.g. ITU-T I.150 'ATM functional characteristics'). The appendix then provides a short description of the contents of each recommendation and then the references of related specifications (or textually similar documents) issued by other standards bodies including ETSI and ANSI T1 Committee.

5.7 Structure of ITU-T Recommendations on ATM

A general understanding of the principles and aims of ATM and B-ISDN is crucial to being able to comprehend the standards. (Having read this book, the reader will be much better prepared than beforehand.) The first standards which should be read are ITU-T recommendations I.150, I.311, I.113 and I.361. These cover the basic technical principles of ATM, the structure of the ATM cell and the use of the ATM Adaptation Layer (AAL) as the means by which different services and service classes can be converted into a format suitable for carriage by an ATM network.

Recommendations in ITU-T's *I-series* are all related to ISDN (Integrated Services Digital Networks). These recommendations define the architecture and functionality of such networks as well as the various interfaces and protocols employed by network devices when communicating with one another. Those relating to reference models are generally overview recommendations, defining network reference points and related terminology.

Recommendations relating to *UNI (user–network interface)* define standard physical and communication interfaces which should be used between a user's end equipment and the network itself. Standards relating to *NNI (network–node interface or network–network interface)* define the interconnection of the various network elements (nodes) within an network. *inter-network interface (INI)* or *inter-carrier interface (ICI)* is defined to regulate the measures necessary for connecting networks which are operated by different operators.

Recommendations in ITU-T's *G-series* define transmission equipment. With regard to ATM, the G-series recommendations relating to *SDH (synchronous digital hierarchy)* and to monomode optical fibre transmission are cross-referenced with certain of the ATM recommendations in the I-series.

Recommendations in ITU-T's *Q-series* define the signalling and protocols employed between devices within the network for indicating the establishment of connections or for managing and controlling the network. Thus, for example, Q.2931 describes the protocol by which an end equipment requests the network to set up a connection to a particular destination port provided within SDH.

6 The ATM Network Reference Model

This chapter presents the ATM network reference model. The model defines the different types of network elements which make up an ATM network and the standard interfaces which should be used to interconnect them. Both the UNI (user–network interface) and NNI (network–node interface) are explained, as are the various different types of communication which may be supported by them (user information transfer, network control and network management). The chapter presents and discusses the various network interfaces, comparing and contrasting the ITU-T network reference model with the extensions of the model and slight differences proposed by ATM forum.

6.1 ATM Network Reference Model of ITU-T

ITU-T's basic network reference model is illustrated in Figure 6.1. It defines three basic interfaces:

- UNI (User–network interface)
- NNI (Network–node interface)
- INI (Inter-network interface)

UNI provides a standard for connection of customer equipment to a (presumed public) ATM network, while INI allows for the interconnection of public ATM networks (in the same way that public telephone network operators would interconnect to convey international traffic). NNI is an internal network interface within the network operator's domain which allows him to buy ATM network components from different manufacturers

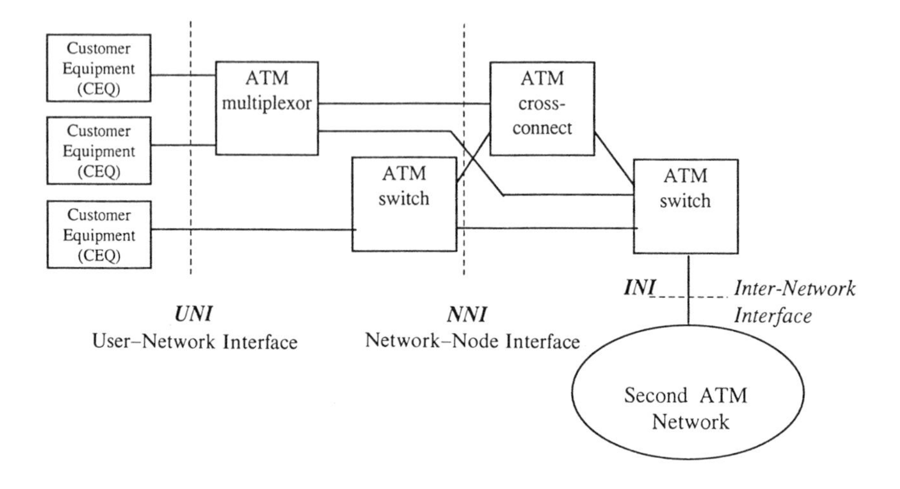

Figure 6.1 The components of an ATM network

and interconnect them to make a network. It is interesting to note the particular effort made to distinguish the different types of network node (ATM multiplexor, ATM crossconnect and ATM switch). These terms reflect terminology carried over from the world of public telephone networks.

6.2 ATM Forum Network Reference Model

The network reference model of ATM Forum (Figure 6.2) extends that of ITU-T by taking care to distinguish between the *private* and *public* parts of an ATM network. Thus the model caters for the connection of *private* corporate and campus ATM networks to a *public* ATM network.

It is interesting to note in Figure 6.2 that ATM Forum depicts the interfaces UNI and NNI between sub-networks and not between individual nodes. This reflects the philosophy particularly of data network equipment manufacturers who have traditionally used 'proprietary' interfaces between the nodes within a sub-network to enable them to differentiate their products from those of their competitors by offering improved network management and service features. It is a good example of the slight difference in emphasis between ATM forum and ITU-T.

The ATM Forum network model distinguishes between the *public UNI* (which it also refers to synonymously as the *public network interface*) and the *private UNI* (synonymously called the *private local interface*). In addition, ATM Forum defines two specific sub-types of the NNI, giving them the names

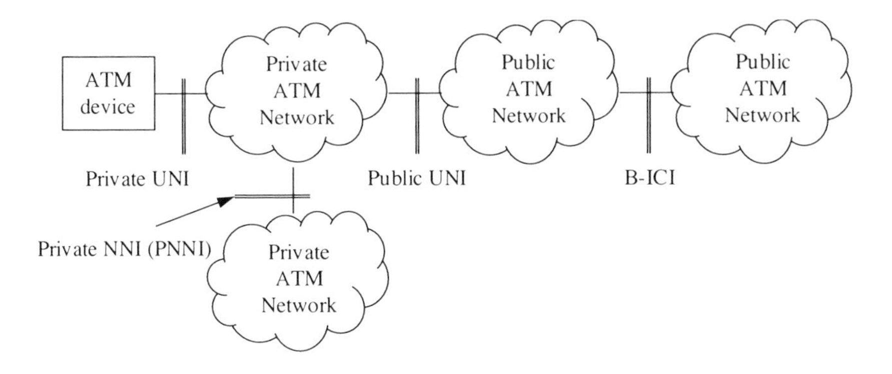

Figure 6.2 ATM Forum network reference model

PNNI (private network–node interface or *private network–network interface)* and *B-ICI (broadband inter-carrier interface)*. At a basic level, these both use the NNI. They differ in the types of network management and administration possible over them.

The ATM Forum UNI is the best specified of all interfaces available. Two versions are prevalent in currently available ATM switches. These are versions 3.0 (UNI v3.0) and 3.1 (UNI v3.1). In their time they were both closely aligned with concurrent ITU-T standards. Version 3.1 incorporates the ITU-T recommendation Q.2931 for call setup of switched virtual circuits (SVCs). Version 3.0 is based on an earlier draft of Q.2931 and is incompatible with v3.1. These specifications are published by Prentice Hall. [see footnote p. 78]

PNNI includes a number of functions for *discovery* of the network topology and for optimal routing through it. This is well suited, for example, to a university or large campus private ATM network in which individual departments may be responsible for adding new switches and end-user devices to the network on their own initiative. The PNNI *topology state* functions enable other switch nodes within the network to keep abreast of the topology and connected devices.

B-ICI, in contrast to the PNNI, defines a more secure interface – reinforcing the organizational boundary between different public network operators. The B-ICI interface adds specific functions to the standard NNI to allow for *contracting traffic,* and for monitoring and management of an *interconnect* between different operators' networks (e.g. in the USA the *interconnect* between *LECs (local exchange carriers)* and *IECs (inter-exchange carriers)*). The B-ICI interface also supports the capabilities necessary for transitting networks (like IEC networks), and allows for the support of *cell relay service (CRS), frame relay service (FRS)* and *SMDS (switched multimegabit data service)*. The B-ICI interface is thus an important precursor to the regulated interconnection of public ATM networks, at least in the USA.

ATM Forum has defined three *interim* interfaces for use at the PNNI, B-ICI and for management control of the network. These are, respectively,

- *IISP (interim inter-switch signalling protocol* – a forerunner of the PNNI signalling protocol based on ATM forum UNI v3.1*)*;
- *BISSI (broadband inter-switching system interface* – a forerunner of the NNI signalling protocol for use in and between public ATM networks*)*; and
- *ILMI (interim local management interface* – a protocol for management of ATM network devices based on the Internet *simple network management protocol, SNMP).*

ITU-T has not recognized these interim standards.

6.3 The *User-Network Interface (*UNI*)*

The *User–Network Interface* is the standardized interface for connecting users' end devices to an ATM network for the purpose of communication. ITU-T recommendation I.413 defines the *reference points* of the UNI. Actually, the various *reference points* are a range of slightly different interfaces all of which are UNI interfaces, but which differ according to how the device is connected to the network (Figure 6.3).

The TB-interface is the basic UNI interface by which user equipments are connected to an ATM network at the B-NT1. The *B-NT1 (broadband ISDN network termination type 1)* provides line terminating functions for the public network operator (power feeding to line, management test functions, etc.).

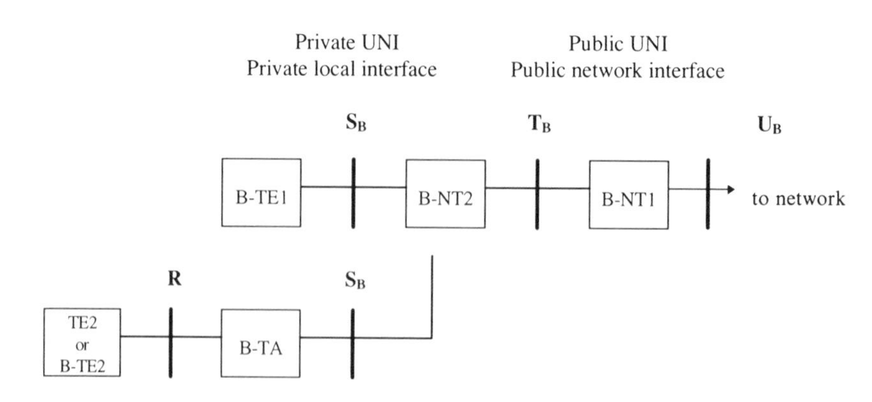

Figure 6.3 B-ISDN reference configurations at the UNI

Alternatively, the user device can be connected directly to the line (UB-interface). Indeed, in the case where the optical version of the UNI is employed by the customer's equipment (CEQ), there may be no *B-NT1*. A further possibility, due to the regulatory requirement in the United States that *customer premises equipment (CPE)* including NT-devices are not provided by public network operators (so that they are open to competition in their supply), is that the public network operator provides a UB-interface.

Where an electrical interface is used at the UNI, a B-NT1 may be required to perform electrical to optical conversion of the signal for connection to a fibre connection. Alternatively, the ATM network operator may choose to install a device to provide for network monitoring or management functions. If so, this device is also classified as a B-NT1, and the user side interface to it is then the UNI, reference TB. The TB-interface is equivalent to the T-interface of narrowband ISDN. The B-suffix is merely to denote its connection with B-ISDN.

B-NT2 (broadband ISDN network termination 2) is the notation used in the ITU-T standards to denote some form of user–provided switching device, which permits a multiple number of end-user devices to share the same network connection. Thus a B-NT2 has one TB-connection, but may have multiple SB-ports, allowing multiple user terminal equipments to be connected to it. The equivalent in a narrowband ISDN is an NT2. An example of an NT2 device is an office PBX–an office-installed telephone exchange which switches a large number of individual extensions and a small number of main exchange lines. In this way better use can be made of the main exchange lines. In reality, the SB-interface will be very similar (if not identical) to the TB-interface, so that end terminal equipments (B-TEs) can be connected directly to the ATM network (TB-interface).

B-TE1 (broadband ISDN terminal equipment type 1) is the notation used in the ITU-T standards to denote customer equipment (CEQ) which has been developed to operate over ATM networks. It is an ATM-compatible end device.

A *B-TA* is a *broadband ISDN terminal adaptor*. This is a device provided to convert signals from non-ATM-compatible end devices (*B-TE2*) into a format suitable for carriage over an ATM network. Thus the R-reference point is merely a notation used within the ATM standards texts to refer to other types of end-user device interfaces (i.e. any interface not covered by ATM specifications).

ITU-T Recommendation I.413 also defines a W-reference point in conjunction with a *medium adaptor (MA)*. A number of medium adaptors perform the function of a B-NT2, allowing multiple end devices to be connected to a single UNI network connection. This is a *multipoint application* (Figure 6.4). The W-reference point is to be found between MAs. The technical realization of the W-reference point may be a non-standard (i.e. manufacturer proprietary) interface.

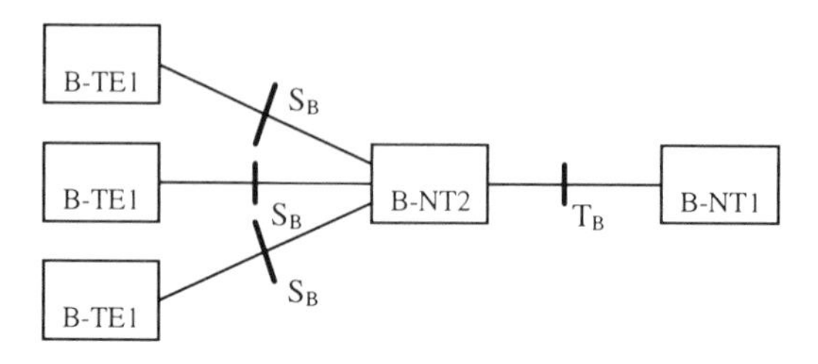

(a) Cetnralized (hub-type) B-NT2 configuration

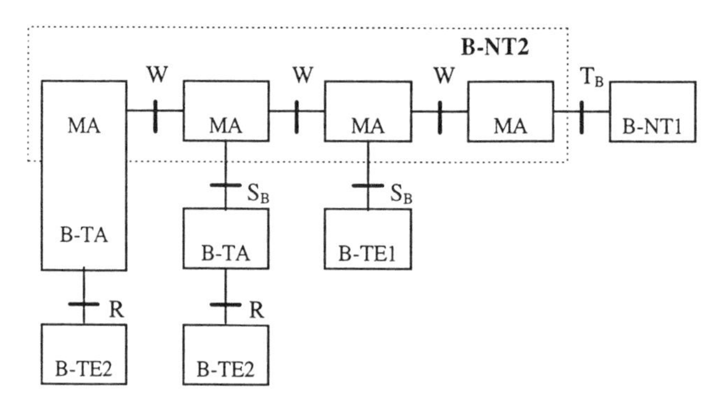

(b) Bus-type multipoint cofiguration

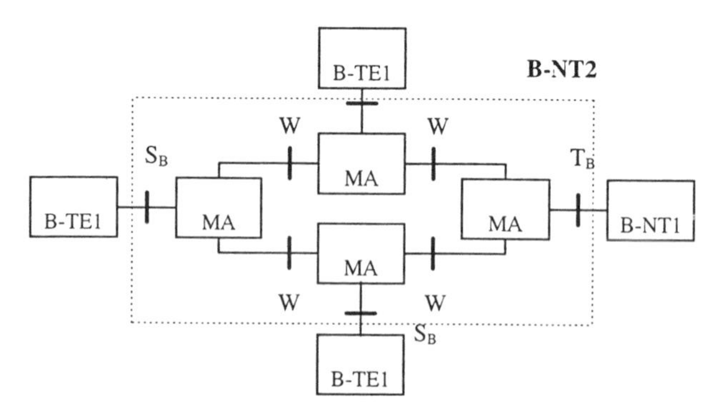

(c) Ring-type multipoint configuration

Figure 6.4 Multipoint configurations of the ATM User-Network Interface (UNI)

The various configurations appearing in Figure 6.4 reflect some of the complexity of the standards and the extensive capability of ATM technology. But do not let yourself be put off by the mass of jargon before you observe that the wiring configurations of the ATM UNI mirror the classical in-house wiring configurations of customer premises. The hub-type configuration (Figure 6.4(a)) is typical of a structured wiring system centred on a wiring cabinet per office floor. The hub topology is also common in *local area networks (LANs)* based upon LAN *hubs*. The bus topology (Figure 6.4(b)) is like that of a classical coaxial cable Ethernet LAN. The ring topology (Figure 6.4(c)) is like that of a Token Ring LAN or an FDDI network. Whether Medium Adaptation devices of these various types are actually developed by manufacturers remains to be seen, but we can at least note that the standards lay a basis for the usage of some existing infrastructure.

ITU-T recommendation I.414 provides an overview of the detailed recommendations which specify the physical interfaces relevant to the various UNI reference points.

6.4 Communication between User and Network Elements – the User, Control and Management Planes

Before two *customer equipments (CEQ)* may communicate with one another (i.e. transfer information) across the *user plane (U-plane)* of an ATM network, a connection must first be established. The connection is established by means of a *control* or a *management* communication between the CEQ and the network. This communication may take one of five forms (as defined by ITU-T recommendation I.311):

- *Control plane communication (access)*
- *Control plane communication (network)*
- *Management plane communication type 1*
- *Management plane communication type 2*
- *Management plane communication type 3*

These various types of communication control and manage the various network elements of an ATM network (VP (virtual path) crossconnect, VC (virtual channel) crossconnect, VP–VC crossconnect, VP switch, VC switch, VP–VC switch) as illustrated by Figure 6.5.

A *control plane communication (access)* is one conducted between CEQ and a VC switch in the ATM network. During such a communication (UNI

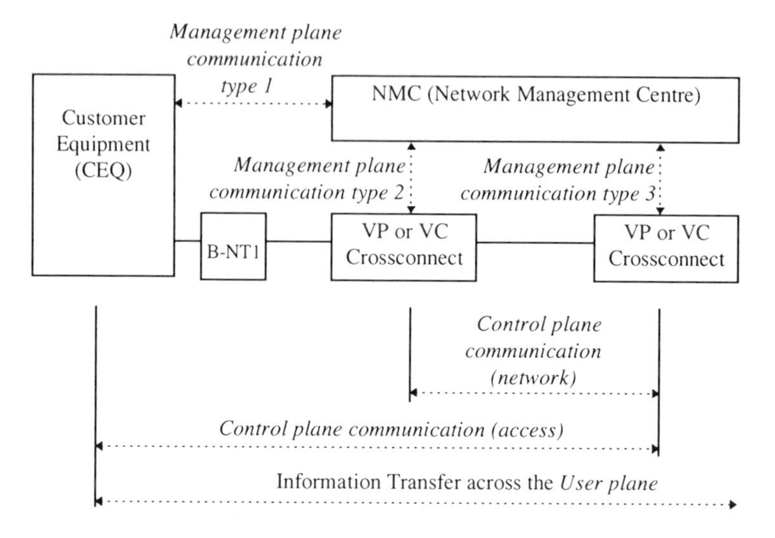

Figure 6.5 User, control and management planes of an ATM network (ITU-T)

signalling), the network user is able to establish or release connections (SVCs, switched virtual circuits) across the ATM network much like dialling a telephone number in a telephone network. *Control plane communications (network)* will follow, as the ATM network switch communicates (network signalling) with other nodes in the network to establish the complete network connection. Once established the user is said to *transfer information* (i.e. communicate) across the *user plane*.

Alternatively the connection could be established under manual intervention of the service technicians at the network management centre. In this case, the user sends a *management plane communication type 1* from his CEQ to the NMC to request the establishment of a permanent connection. This could be carried by UNI signalling or could simply be a telephone call. The various network elements (VP, VC crossconnects) are then configured from the NMC by means of messages sent by *management plane communication type 2.*

Management plane communication type 3 is initiated by ATM VC switches which, on receipt of request for connection set-up by means of *control plane communication (access)*, require to refer to the NMC for information or authority before proceeding with the setup. (It could be the case, for example, that certain high bandwidth connections be subject to NMC authority in order to prevent network congestion at peak times.)

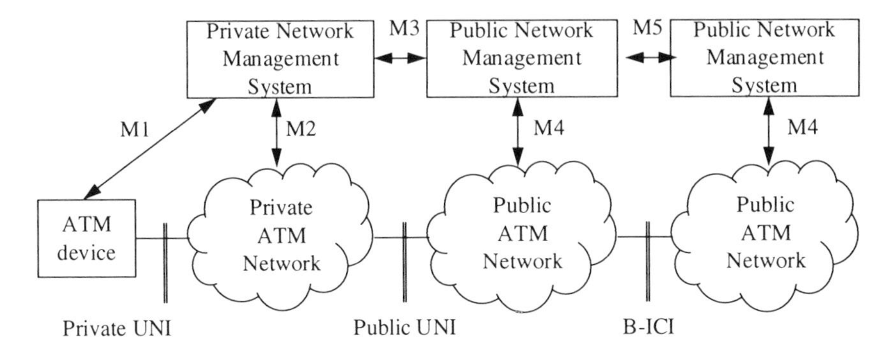

Figure 6.6 ATM Forum network management reference model

6.5 ATM Forum Network Management Model

As with the ATM Forum network interface reference model, the ATM Forum network management model (Figure 6.6) differs from that of ITU-T recommendation I.311 by the strict separation of public and private ATM network components, though the basic principles are the same.

Confusingly, ATM Forum refers to M-interfaces rather than to *management plane* types, and uses different numbering (ATM Forum interface M1 is equivalent to ITU-T management plane type 1, but ATM Forum interface M3 is not equivalent to ITU-T's management plane 3).

ATM Forum's M3- and M4-interfaces are already extensively defined by ATM Forum, respectively for *customer network management (CNM) of public ATM network service* and for network management of public ATM networks. The other interfaces are less well defined.

6.6 Setting Up ATM Connections – Signalling to an ATM Network at the UNI

Both management and control communication in an ATM Network take place via *signalling virtual channels (SVCs)*. A *meta signalling virtual channel (MSVC)* is always available at every UNI. This is a signalling channel of a fixed bandwidth to be found in the virtual path with VPI = 0, and has a network standardized VCI value. The MSVC is used to establish *signalling virtual channels (SVCs)*. It is via the SVCs that the user signals his management or control plane needs.

By means of the *meta signalling virtual channel*, the end device (CEQ) can establish either a SVC (signalling VC) to the VC switch (for control plane communication (access)) or to the network management centre for management plane communication. A *service profile identifier (SPID)* carried in the meta-signalling helps the network determine which service the user wishes to use, and enables a signalling virtual channel to the appropriate *signalling point* server to be established. (Meta signalling is defined by ITU-T recommendation Q.2120.)

Using the SVC (signalling VC), virtual channel connections of SVC-type (switched virtual circuits) may be established for *user plane communication* across the ATM network. The exact signalling procedure used in conjunction with setting up *SVCs* (*switched virtual circuits*) may be one of two types, either:

- user-to-network signalling (e.g. for establishing or releasing a VCC for end-to-end user plane communication); or
- user-to-user signalling (e.g. for establishing or releasing a VCC within a pre-established VPC between two UNIs).

The exact form and information content of the signalling message is defined by ITU-T recommendation Q.2931.

Alternatively, a user plane VCC could be established without requirement for signalling where the user subscribes to a permanent connection (*PVC* or *permanent virtual circuit*).

The value of the VCI (virtual channel identifier) allocated to the VCC set up across the UNI is assigned either by the network, by the user, by signalling negotiation or by standard agreement.

6.7 Signalling at the NNI

Signalling virtual channels are also used for intra- or inter-signalling requirements (i.e. for control or management plane communications) between the network elements of an ATM network, but meta signalling is not used to assign the signalling channels. Instead one VCI is reserved per VP as a permanent signalling channel.

The signalling in B-ISDN networks (including ATM) is designed in a very similar way to the SS7 signalling networks used in *narrowband ISDN (N-ISDN)*. Signalling takes place between *signalling points (SPs)*, and may be conducted either in *associated mode* or in *quasi-associated mode*.

The *signalling points*, as we saw briefly in chapter 2, are functionality built into ATM switches associated with the control and switching of the service requested by the user. Thus a signalling point in the case of a voice telephone service would be an ATM switch performing switching of voice connections. In order to perform the switching, the switch needs to be capable of analysing the dialled telephone number (or equivalent), decide to which next exchange the call should be routed and signal to the *signalling point* in this next exchange. Such is an example of *associated mode signalling*.

In *quasi-associated mode signalling*, the signalling information is directed to the *signalling point* via a route which is different from that which the call or connection itself will take (once switched through). In *quasi-associated* signalling, the signalling information passes over a specialist *signalling network* via transit exchanges called *signalling transfer points.* Figure 6.7 illustrates the concepts of *signalling points (SPs)* and *signalling transfer points (STPs).*

Operating in the *associated mode*, node A of Figure 6.7 would signal directly to the SP at node B and then switch the connection to follow the same path. In the *quasi-associated mode*, node A signals to node B to expect the connection, but sends the signalling information via the STP at node C and switches the connection over the direct path to node B. The advantage of quasi-associated signalling is that it can be used to dramatically reduce the total number of signalling links in the network, thereby much reducing the cost of the signalling point hardware needed at each exchange. This is achieved by connecting SPs in star-connected fashion to a small number of STPs.

A further advantage of a signalling network architecture which is similar to that of SS7 signalling used in N-ISDN is that it lays the basis for interconnection of B-ISDN and ATM networks to the *intelligent network*. This offers the potential for powerful service control functions to be added to ATM network

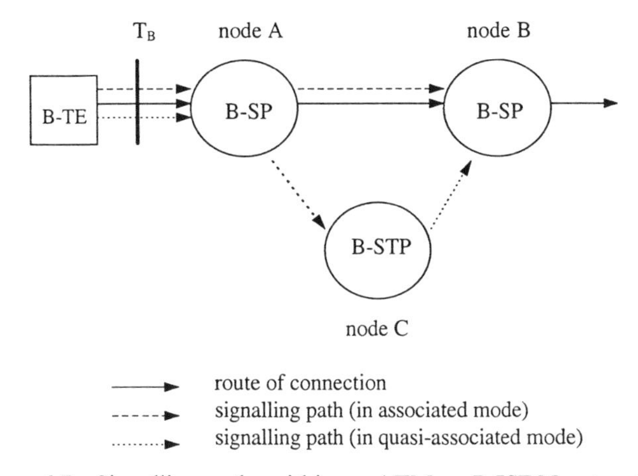

Figure 6.7 Signalling paths within an ATM or B-ISDN network

services, as we discussed in chapter 4. Thus the B-ISDN architecture includes the intelligent network functions *service switching point (SSP)* and *service control point (SCP)*. A broadband SSP is an ATM switch with functionality added to it to be able to converse with and *query* a central database *(SCP)* for instructions on how to handle a particular service or a particular customer's call.

The strength of the *intelligent network* is the ability to control the services in the network from a single database. The software and configuration information need only be created once (rather than duplicated in every node). This leads to much greater consistency in the handling of services and the potential to realize new services much more quickly.

6.8 Operations and Maintenance Virtual Connections (OAM VCs)

Operations and Maintenance VCs are established within an ATM network in order to allow the network management centre personnel to monitor, reconfigure and maintain the network. These may be established at either UNI or NNI. The functions made available by OAM VCs are discussed in chapter 10.

UNI v4.0

ATM Forum's UNI v4.0 is currently being heralded as *the* UNI standard which will start to realise the full potential of ATM. It concentrates on the AAL1 and AAL5 adaptation layer types, substituting. AAL2 by dividing AAL1 into *real time (RT)* and *non-real time (NRT)* services.

7

The ATM Protocol Reference Model

As with many modern telecommunications technologies, ATM is based upon a number of carefully defined techniques and procedures (protocols) arranged to cope with the task of reliable information transfer across the network. The set of different functions of ATM and the means by which they interact with one another is defined by the ATM protocol reference model. The different functions defined in the various layers of the model each have clearly defined purposes ranging from the control of access to the network, the establishment of connections and the transfer of information, to the detection and correction of network problems. In this chapter we present the ATM protocol reference model, and discuss the various 'planes' and 'layers'(functions) defined by it.

7.1 The Principles of a Protocol Reference Model

Defining the procedures and functions to be performed by a network in a layered manner eases the understanding of the procedures (*protocols*) required for reliable information carriage across it. This has three main benefits. First, complex techniques may be more easily broken down into individual layers, each of which can then be specified by a different specialist or group of specialists. The subsequent development of hardware and software to perform the function may similarly be carried out and tested in manageable chunks. Second, in operation, problems or faults can be more easily traced to one of the individual functional layers. Thirdly, the technology can be more quickly understood by newcomers.

A *protocol reference model* is usually defined in terms of a number of layers. Each layer provides a defined *service* to the layer above it. The exact nature of the service provided is controlled by *primitives* (defined commands) issued by the layer above it. The service is said to be delivered at the *service access point (SAP)*. This is an imaginary point between the functional layers.

7.2 The B-ISDN/ATM Protocol Layers

Table 7.1 lists the protocol layers of ATM. We use this in the discussion which follows to explain the relationship of the various layers to one another, and also to explain some of the complicated terminology used in the specifications of the individual protocols.

7.3 The ATM Transport Network

The foundation of the various protocol layers (the *protocol stack* – the set of functions which together make information transfer possible) is the physical medium used for the carriage of electrical or optical signals. The *physical layer* is a specification which defines exactly the medium that should be used – what electrical or optical signals and voltages, etc., should be used. In addition it sets out a procedure for transferring data across the line, providing for clocking of the bits sent and the monitoring of the equipment. The physical layer of ATM is similar in function to the physical layer (layer 1) of the *open systems interconnection (OSI)* protocol stack (the ISO and ITU-T standard protocol stack for telecommunications networks).

Table 7.1 The functional layers of ATM

Layer name	Sublayer name	Further sublayer
Higher layers	Convergence Sublayer (CS)	Service Specific (SS)
ATM Adaptation Layer (AAL)		Common Part (CP)
	Segmentation and Reassembly (SAR) Sublayer	
ATM Layer	VC Level	
	VP Level	
Physical Layer	Transmisssion Convergence (TC) Sublayer	
	Physical Medium	

• Service Access Points (SAP)

The physical layer is divided into two sublayers. These are the *physical medium* sublayer and the *transmission convergence (TC) sublayer*. We discuss these in depth in chapter 8. The physical medium sublayer defines the exact electrical and optical interface, the line code and the bit timing. The TC sublayer provides for framing of cells, for cell delineation, for cell rate adaptation to the information carriage capacity of the line, and for operational monitoring of the various line components (*regenerator section (RS), digital section (DS)* or *transmission path (TP)*).

The *service access point (SAP)* of the physical layer *(PL-SAP)* is the conceptual point shown in Table 7.1 where the physical layer *service* is delivered to the ATM layer above it. The ATM layer controls the transport of cells across the ATM network, setting *up virtual channel connections* and controlling the submission rate (generic flow control) of cells from user equipment.

The service provided to the ATM layer by the physical layer is the physical transport of a valid flow of cells (this flow of cells is correctly called a *service data unit (SDU)*, in fact the *TC-SDU (transmission convergence sublayer service data unit)*). The ATM layer controls the service provided to it by means of *service primitive* commands. These are standardized requests and commands exchanged between a function within the ATM layer called the *ATM layer entity (ATM-LE)* and the *TC sublayer entity (TC-SLE)*. They allow, for example, a particular ATM-LE to request the transfer of a flow of cells (service data unit), to the physical layer the opportunity to halt a transfer due to a problem with the physical medium.

The transmission convergence sublayer (of the physical layer) receives information in the form of cells (the service data unit (SDU)) provided to it by the ATM layer. These cells are supplemented by further information, including PL-cells (physical layer cells) and OAM cells (operations and maintenance cells) as we discuss more in chapter 10. The extra information, an example of *protocol control information (PCI)*, turns the TC-SDU into a *TC-PDU (protocol data unit)*. It ensures the correct transmission of information across the physical medium, including *cell delineation* and *scrambling*. The *TC-PDU* is passed to the physical medium sublayer, where it is called the *PM-SDU (physical medium service data unit)*.

Finally, the *PM-PDU* is passed to the medium itself. The form of the PM-PDU (and thus the conversion performed by the physical medium sublayer) is thus electrically and optically dependent upon the type of medium used. An advantage of this is that only the physical medium sublayer need be altered to accommodate a change of the physical medium. Other hardware and software components (e.g. corresponding to the ATM layer) may be reused.

Together the ATM layer, the physical layer and the physical medium are called an *ATM transport network*. An *ATM transport network* (ATM and physical layers) is alone capable of conveying information between network endpoints. However, in order that the information content carried by an

ATM transport network can be correctly interpreted by the receiver, further higher layer protocols are defined. The most important of these is the *ATM adaptation Layer (AAL)*.

7.4 The ATM Adaptation Layer (AAL)

As the name suggests, the *ATM adaptation layer (AAL)* provides for the conversion of the *higher layer information* provided to it by the *higher layers* into a format suitable for transport by an ATM transport network. The *higher layers* are information, devices or functions of unspecific type which require to communicate across the ATM network. As we learned in chapter 6, the information (*higher layer information*) carried by the ATM network may be either:

- **user information** *(user plane)* of one of a number of different forms (e.g. voice, data, video, etc.) as categorized by the AAL service classes (Table 2.1);
- **control information** *(control plane)* for setting up or clearing SVCs (switched virtual circuits); or
- **network management information** *(management plane)* for monitoring and configuring network elements or for sending requests between network management staff (e.g. for establishing new PVCs, permanent virtual circuits).

Like the other layers, the AAL accepts AAL-SDUs from the higher layers (actually a CS-SDU, convergence sublayer SDU) and passes an AAL-PDU (actually an SAR-PDU) to the ATM layer, where it is known as an ATM-SDU. But unlike the ATM and physical layers a number of different alternative services can be made available to the *higher layers* of the communication stack, thus allowing different types of information to be *adapted* for carriage across a common ATM transport network. It is the ATM adaptation layer which gives ATM networks their capability to transfer all sorts of different information types. It is split into two sublayers, the *convergence sublayer, CS* (where the alignment of the various information types into a common format takes place and division into cells occurs) and the *segmentation and reassembly sublayer, SAR* (where the cells are numbered sequentially to allow reconstruction in the right order at the receiving end).

Higher level information is the name given to the information in its raw form requiring carriage by the ATM network. It is passed as a higher layer PDU to the AAL, where it is known as an AAL-SDU. The AAL performs its defined functions by converting the information as necessary and adding its *protocol control information (PCI)*, which will help the AAL function at the receiving end of the network (its *peer* partner) to perform the reverse conversion. The result is the AAL-PDU. This is passed to the ATM transport network, where it is known as the ATM-SDU.

7.5 Peer-to-peer Communication between ATM Protocol Model Layers

Figure 7.1 illustrates the basic *protocol stack* of ATM. It is drawn as a number of layers in which *peer functions* at either end of the physical connection appear to communicate directly between one another (dotted lines in the diagram). The dotted line communication does not in reality take place directly between the *peer partners*, though it appears to them as if it did. In reality their messages are appended with extra information by each of the underlying layers (to ensure reliable carriage) and are actually transmitted by the physical medium which appears in the diagram at the bottom of the stack.

The steps in the real communication are thus:

1. Higher layer presents information to be carried to the AAL as a higher layer PDU.

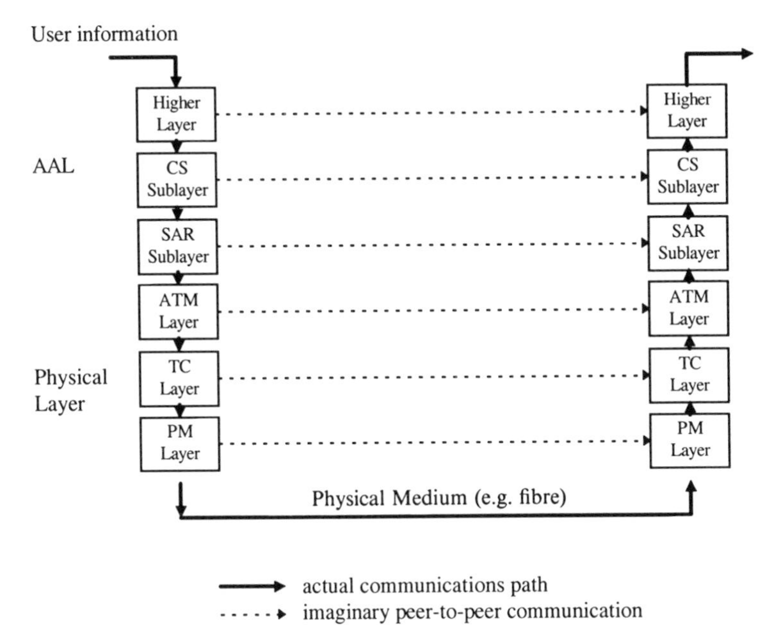

Figure 7.1 Peer-to-peer communication between the functional layers of ATM

2. The higher layer PDU (known by the AAL as the AAL-SDU) is converted by the appropriate AAL functions (according to information type) to the AAL-PDU by adding PCI (protocol control information) and is passed to the ATM layer.

3. The AAL-PDU (known by the ATM layer as the ATM-SDU) is converted by the addition of ATM layer PCI to the ATM-PDU and is passed to the physical layer.

4. The ATM-PDU (known by the physical layer as the PHY-SDU) iconverted by the addition of physical layer PCI to the PHY-PDU and is transmitted over the physical medium.

At the receiving end, everything happens in reverse. The layers are traversed from layer 1 upwards, PDUs becoming SDUs as the various protocol control information (PCI) is removed by each of the layers in turn – reversing the action of its *peer* layer partner at the sending end. It is as if the various layers at sending and receiving ends had talked directly with one another (Figure 7.1).

How the underlying layers performed their functions in order to make this possible need not worry the peer layer functions. Hence the commonly used term *peer-to-peer communication*. The clear functional separation of the layers is what gives the benefits of easy design and operation as we discussed earlier in this section.

7.6 Protocol Stack Representation of End Devices Communicating via an ATM Transport Network

A major benefit of the layered structuring of communications protocols is that intermediate elements within the network do not necessarily need to cope with the functions represented by each of the layers.

Figure 7.2 illustrates the peer-to-peer communications which take place when two user end devices communicate with one another by means of an ATM transport switch. The ATM switch supports only the lowest three protocol layers, and speaks peer-to-peer with each of the ends, relaying information. Meanwhile, at the *ATM adaptation layer (AAL)* and the higher layers, the two end devices communicate peer-to-peer directly over the connection established by the lower three layers.

If we were to monitor the wire between either of the end devices and the ATM switch of Figure 7.2 then we would observe communication at each of the layers. What we actually observe are cells, but structured a little like a Russian doll. The smallest doll right inside is the information we want to carry

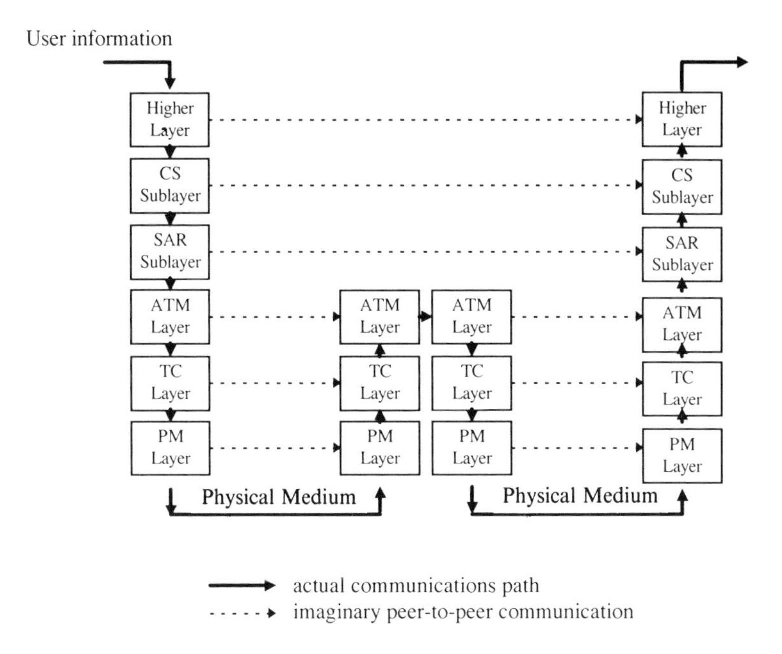

Figure 7.2 Protocol Layer representation of two end devices communicating via ATM Layer switch

between the users (the *higher layer information*). All the other cladding dolls are the protocol information (PCI) – one doll for each of the lower layers, each providing a function critical to the reliable end-to-end carriage of information and its correct interpretation on receipt.

7.7 The User and Control Planes

In chapter 6 we discussed the concept of the *user* and *control planes*. In short, when an end-user device wishes to establish (i.e. dial-up) a new connection across an ATM network then it must signal to the network its desire to do so (equivalent to dialling a telephone number in a telephone network). This signalling takes place on the *control plane*. Subsequently, when the connection to the other end has been established, then the two end-user devices communicate with one another on the *user plane*.

The protocols at the ATM adaptation layer and at the higher layers differ according to whether the device is communicating on the *control plane* or on the *user plane*. The difference arises from the different nature of the information to be carried (the *higher layer information*) in the two cases. In one case (on the *control plane*) the network itself must interpret the higher layer information and react to it. In the case of communication across the *user plane*, the higher layer information may take any number of different forms (speech, data, etc.), but the individual network elements themselves (e.g. switches) may be incapable of recognizing and interpreting these various forms.

The AAL protocols used on the *control plane* will usually differ from the AAL protocols used on the *user plane* of the same connection. The *control plane* AAL (for the user-network signalling in setting up a switched virtual circuit, SVC) will typically need to be suited for data information transfer, while the *user plane* AAL may be required to convey telephone or video signals. For native ATM devices (e.g. the *broadband connection-oriented bearer, BCOB-X)* a *null* AAL may be used (i.e. no AAL is used). Conversely, for PVCs (permanent virtual circuits) set up by network management commands, the *control plane* AAL may not be necessary. (In fact, in the early days of ATM standards no control plane AALs were defined, so only PVCs could be supported.)

Taking account of the difference in protocols used at the ATM adaptation layer and higher layers, but of the commonality at ATM and physical layers, the B-ISDN protocol reference model illustrates the control and user planes as shown in Figure 7.3.

In real switches and ATM end-user devices, common or duplicate hardware and software may thus be used for both control and user planes at ATM and

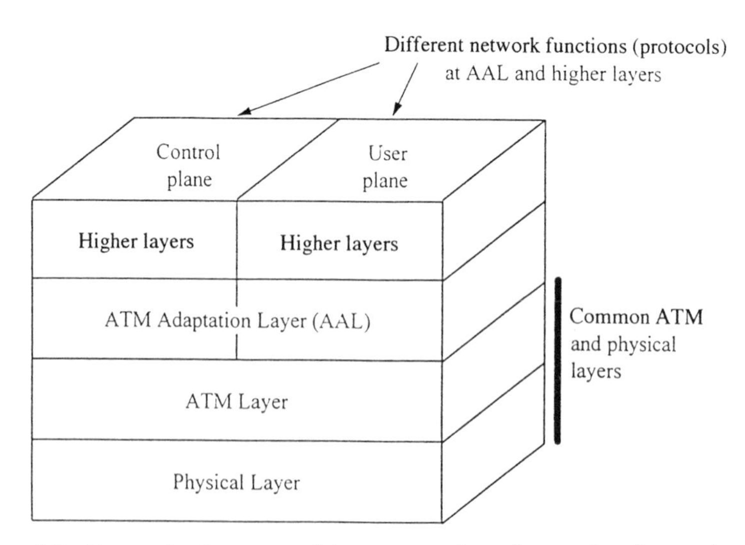

Figure 7.3 Protocol reference model representation of control and user planes

physical layers, but distinct hardware and software will be necessary for signalling and user information transfer at AAL and higher layers.

7.8 Service-specific AAL Control Plane Functions and Protocols

A number of different types of services may be carried by the AAL. In the jargon these are *service-specific convergence services.* Examples of specific *user plane* services offered by the *ATM adaptation layer (AAL)* to the higher layers are:

- Frame relay *SSCS (Service specific convergence sublayer)* service
- SMDS (*Switched multimegabit data service)* SSCS service
- Reliable data delivery SSCS service (a packet-network-like data network service)
- LAN emulation SSCS service
- Desktop quality video SSCS service
- Entertainment quality video SSCS service

At the higher layers of the user plane, further protocols are in development at the ATM Forum, particularly for encapsulating existing telecommunications protocols for carriage by an ATM network.

The exact protocols used at the AAL and higher layers of the *control plane* for setting up ATM SVCs (switched virtual circuits) are based on the AAL Common part for AAL type 5 (see chapter 9), together with a service specific control plane protocol known as the signalling AAL (SAAL). These are a set of protocols defined by ITU-T recommendations Q.2110, Q.2130 and Q.2140. These carry the user-to-network signalling. This is the dialled number information (equivalent to telephone numbers) and network control signals (akin to telephone ringing and ring tone). The combination of protocols used at the UNI are are called *DSS2 (digital subscriber signalling system version 2*) and *MTP3 (message transfer protocol layer 3)*. At the NNI, *B-ISUP (broadband integrated services user part)* and *MTP3* are used. Figure 7.4 illustrates the control layer protocols defined for the UNI and the NNI.

The higher layer protocols of the control plane (DSS2 and B-ISUP/MTP) are equivalent to the DSS and ISUP/MTP signalling protocols of narrowband ISDN. These are the signalling systems used respectively between an ISDN telephone and the first network exchange and between nodes in the network. Both ISUP and MTP are parts of *signalling system number seven (SS7)*. As their names suggest DSS2 and B-ISUP are based heavily upon their narrowband ISDN counterparts, and indeed as we shall see in chapter 8, the numbering

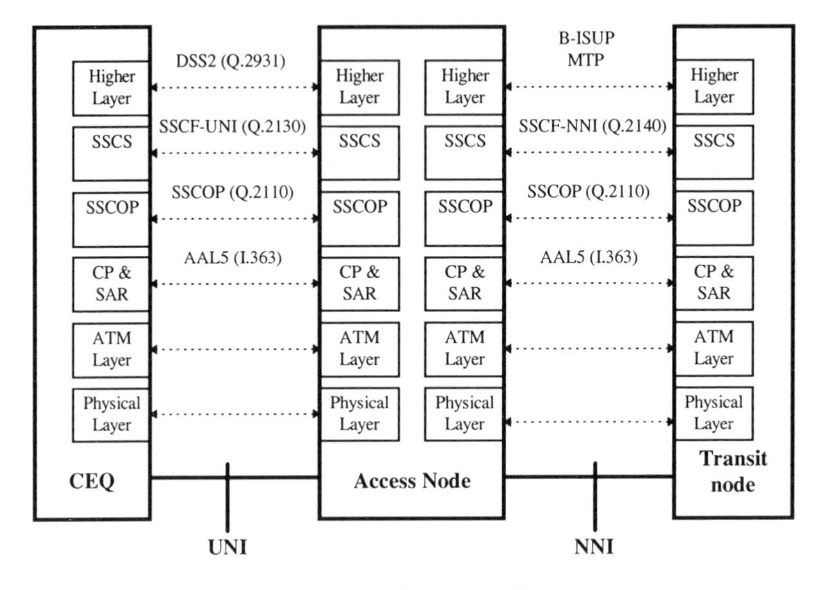

DSS2 = Digital subscriber signalling system 2
B-ISUP = Broadband integrated services user part
MTP3 = Message transfer protocol layer 3
SSCS = Service specific convergence sublayer
SSCF = Service specific coordination function
SSCOP = Service specific connection-oriented protocol
CP = Common part (convergence sublayer)
SAR = Segmentation and reassembly sublayer
AAL5 = AAL service type 5
UNI = User –network interface
NNI = Network–node interface

Figure 7.4 UNI and NNI protocols used on the control plane

plan used in B-ISDN is based on ITU-T recommendation E.164 (i.e. that used in modern digital telephone and ISDN networks). The *signalling AAL* (*SAAL* – consisting of SSCF/SSCOP) allows these protocols to be carried across an ATM transport network in association with an AAL5 service. The SSCOP is quite a complex protocol similar in function to HDLC (higher layer datalink control – OSI layer 2). It provides for guaranteed sequence integrity, error detection and correction and flow control of data blocks between endpoints of the ATM connection. The SSCF provides an additional functionality which in effect simplifies the use of SSCOP when used in the specific case of network signalling.

B-ISUP (Broadband integrated services user part) is defined by ITU-T recommendations Q.2761–Q.2764. The message transfer protocol is identical to that of SS7 (signalling system 7 – the signalling system used between switches in the narrowband ISDN). It is defined by ITU-T recommendations Q.704, Q.707 and Q.782.

7.9 The Management Plane

Management plane protocols are those used for the management of the network. The management plane itself is split into *layer management* and *plane management*.

An example of *layer management* is the meta-signalling procedure we discussed in chapter 5. It is an administrational procedure of the network itself, used to set up signalling connections between customers' end-user equipment (CEQ) and signalling points (SP) in the network for *control plane communication*. Layer management also handles the operation and maintenance (OAM) information flows as specific to the layer concerned. Layer management is defined by ITU-T recommendation Q.940.

Plane management performs the management of the network as a whole, coordinating the actions of the layer management, control and user planes. This leads us to the final standard representation of the B-ISDN protocol reference model as it appears in the cubic form of Figure 7.5.

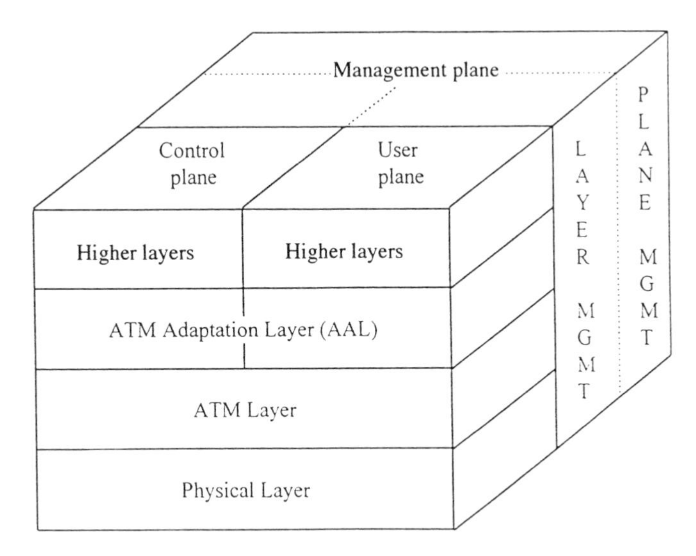

Figure 7.5 The B-ISDN protocol reference model

8

The ATM Transport Network (the ATM and Physical Layers)

The combination of ATM Layer and Physical Layer is an ATM transport network. This is a highly flexible and powerful type of telecommunications network based on a statistical multiplexing technique known as cell switching. This chapter consolidates the various aspects of the ATM and physical layers about which we learned in earlier chapters to give a complete picture of an ATM transport network. We discuss first the structure of the ATM layer, the types of connections and network elements, the setting-up of connections and the maintenance of acceptable connection quality during transmission. We then go on to discuss the range of physical interfaces and media over which ATM cells may be transported.

8.1 The Structure of an ATM Transport Network

ATM transport networks provide for the switching of connections on an on-demand or permanent basis between end-user ports to which *customer equipment (CEQ)* is attached. The connection which is established between the CEQs is correctly called a *virtual channel connection (VCC)*, the endpoints of which (where the CEQ are attached) are called *virtual channel connection endpoints (VCCE)*.

As we learned in chapter 2, a VCC is made up by concatenating *virtual paths (VPs)*, which themselves may traverse more than one physical *transmission path*. In turn, a transmission path may be made up of more than one *digital section*, comprising optical fibre, radio or other physical transmission means.

In Figure 8.1 we illustrate a possible virtual channel connection, made up of

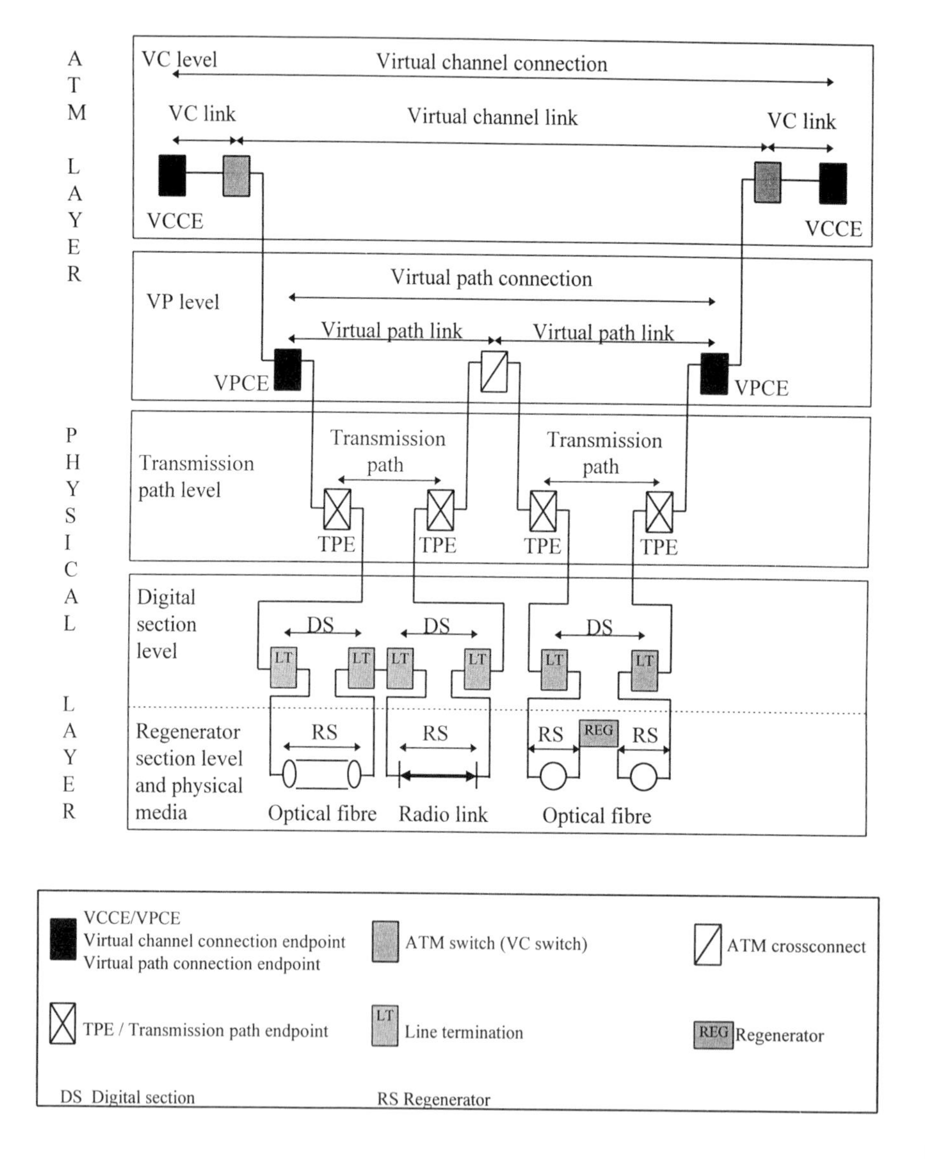

Figure 8.1 Hierarchical structure of an ATM transport network

three concatenated virtual channel links (each a *virtual path connection, VPC*). The diagram shows how the middle virtual path is made up of two separate *transmission paths (TPs)*, one composed of an optical fibre *digital section (DS)* and the other composed of two *digital sections*, one being optical fibre, the other digital radio. The diagram further illustrates how ATM standards also allow a digital section to be broken down into separate *regenerator sections (RSs)*. The diagram demonstrates the different layers (*ATM layer* and *physical layer*) of an ATM transport network and summarizes the associated terminology, demonstrating the interrelationship between the different network components.

The nodes which go to make up an ATM network are switches (VC switches) which switch together virtual channel links (*virtual paths*) to create VCCs. These are devices intended to switch on a call-by-call basis, according to the dialled number (like the exchanges of a telephone network). Crossconnects (VP crossconnects) are devices more akin to patch panels or distribution frames. They crossconnect virtual paths but do not react on a call-by-call basis and are thus relatively simple and cheap devices. ATM switches and crossconnects are also illustrated by Figure 8.1.

Figure 8.2 also illustrates the relationship between virtual connections, virtual paths and transmission paths in another manner. In particular, Figure 8.2 shows how virtual paths are to be found 'inside' transmission paths, and virtual connections 'inside' virtual paths. Thus a single transmission path may carry many different virtual paths, each of which in its own right may contain multiple virtual connections.

Figure 8.3 illustrates the different functionality of customer equipment (CEQ), ATM switches and ATM crossconnects. The dotted lines of Figure 8.3 represent the virtual channel and virtual path connections. But, while the diagram suggests the virtual paths and virtual connections may be connections in a real sense, they are in reality only a *logical* (or *virtual*) fraction of the total bandwidth of the various transmission paths (as Figure 8.2 demonstrated).

From Figure 8.3 we see clearly how an ATM crossconnect possesses switching functionality within the ATM layer only at the VP level. Meanwhile,

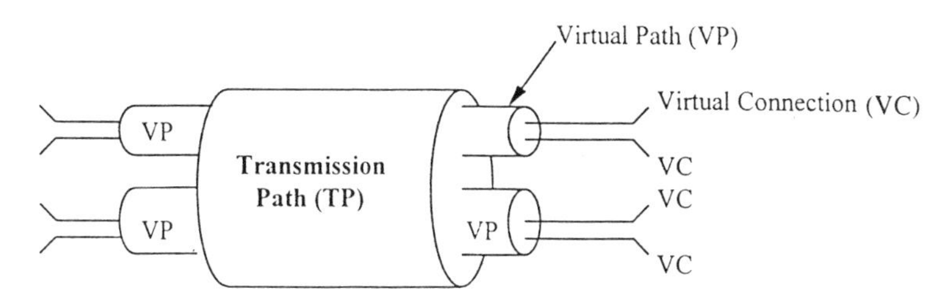

Figure 8.2 The carriage of virtual connections by virtual paths and transmission paths

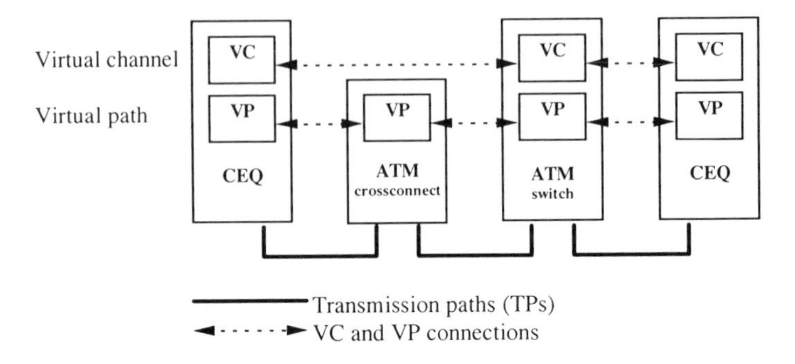

Figure 8.3 ATM connections, switches and crossconnects

it is the VC functionality in the ATM switch and the end devices (CEQ) that deal with virtual channel connections.

As a point of jargon, switching (i.e. crossconnect and switching functionality) is termed in the ATM standards to be a *connection related function (CRF).*

8.2 Functionality of the ATM Layer

The ATM Layer is the crux of ATM switching (and thus of the switching in B-ISDN networks). It is the ATM layer which provides for the cell switching of 53 octet cells. The technical concept of cell switching and its origins in statistical multiplexing we discussed in considerable detail in chapter 2. As a summary, Table 8.1 lists the main functionality provided by the ATM layer. This is the ATM layer *protocol control information (PCI)* coded into the 5-byte cell header.

8.3 Payload Types

The defined ATM layer payload types (as defined by ITU-T I.361) are as shown in Table 8.2.

8.4 ATM Network *Addressing* and Call Establishment

VPIs (virtual path identifers) and VCIs (virtual channel identifiers) are both forms of ATM layer *addresses*. However, it is confusing to think of these fields

Table 8.1 Functionality of the ATM Layer cell header

ATM Layer function (header field)	Acronym	Purpose
Virtual Channel Identifier	VCI	This if the *address* used to identify individual virtual channels within a given physical transmission path. Though the term *address* is the correct terminology, it is easier to think of it as 'label' – in some ways equivalent to the colour-coding of individual copper wires in a multiple copper-wired cable.
Virtual Path Identifier	VPI	This is the *address* used to identify individual virtual paths within a given physical transimission path. Though the term *address* is the correct terminology, it is easier to think of it as a 'label' – in some ways equivalent to the colour-coding of individual copper wires in a multiple copper-wired cable.
Payload Type	PT	The payload type field indicates whether the cell payload (i.e. information field) contains user information or management information. Where it contains management information it additionally indicates what type of OAM (operation and maintainance) cell is included. The value held in the PT-field is called the *payload type identifier (PTI)*.
Generic Flow Control	GFC	Generic flow control is a means of regulating the rate at which cells are submitted to the network by customer end devices (CEQ). When GFC is set to the *controlled mode* (i.e. the value is non-zero) certain types of lower priority traffic may not be submitted by CEQ to the network. This helps to prevent congestion and buffer overflow in multiplexors near the periphery of the network.
Cell Loss Priority	CLP	The cell loss priority supports a mechanism for alleviating instantaneous network congestion caused during statistically infrequent bursts of heavy cell loading. Cells with the cell loss priority bit set '1' shall be discarded in preference to those where the value is set at '0'.
Header Error Control	HEC	The header error control bits are used for detecting errors in the cell header which have arisen during transmission. Where the HEC indicates errors, the cell is said to be *invalid*.

as providing addresses in the sense of addresses as written on envelopes and posted into letterboxes. Instead they are labels, which identify the individual VPs and VCs carried within a given transmission path (like the coloured sheaths of copper wires – they help to identify the connection at both ends of

Table 8.2 Payload types defined at the ATM Layer

Bits (cell header, octet 4)	Payload type identifier value 432	Interpretation
	000	User data cell, congestion nto experienced. ATM-user-to-ATM-user indication = 0. Should this cell tpe be received by a congested node, then the value should be changed to 010, to indicate congestion somewhere along the connection.
	001	User data cell, congestion not experienced. ATM -user-to-ATM-user indication = 1. Should this cell type be received by a congested node, then the value should be changed to 011, to indicate congestion somewhere along the connection.
	010	User data cell, congestion experienced, at some point along the connection. ATM-user-to-ATM-user indication = 0.
	011	User data cell, congestion experienced at some point along the connection. ATM-user-to-ATM-user indication = 1.
	100	OAM (operations and maintaintance) cell, F5 segment associated (see chapter 10).
	101	OAM (operations and maintainance) cell, F5 end-to-end associated (see chapter 10).
	110	Resource management cell.
	111	Reserved for future functions.

each subsection of it). VPI and VCI values are unique at a transmission path level (i.e. for each UNI and NNI), but may be duplicated many times within the network as a whole. (The colour of the sheathing of an electrical circuit might also vary between the various subsections of the total circuit length – viewed end-to-end the same 'wire' may appear red at one end and brown at the other).

A second form of *ATM addressing* is *logical network addressing*. The *logical network address* identifies each port uniquely within the network. It is the *logical network address* which is thus equivalent to a telephone number or an address written on an envelope in the post. Since this information is not carried within the ATM cell header, you might wonder how the network can possibly set up the connection. The answer is straightforward. As we learned in chapters 6 and 7, connections are set up by *signalling* (over the *control plane*) from a *CEQ* to a *signalling point (SP)* by means of a *signalling virtual channel (SVC)*.

The *protocol* used over an SVC (signalling VC) at the *user–network interface (UNI)* is the *digital subscriber signalling system 2 (DSS2)*, as defined by ITU-T recommendation Q.2931. This signalling system allows the CEQ to identify itself by its *logical network address* (*network address* for short) and communicate its desire to establish a connection of a given type to a remote user port bearing a second identified *network address*. Using this information, the network node may allocate a free virtual channel (indicating VPI and VCI values to be used by the CEQ), and thereby connect the CEQ to the network. Meanwhile, the node also determines the best onward *route* for the connection and signals forward to the next exchange using B-ISUP signalling (Figure 8.4).

8.5 ATM Network *Routing*

ATM standards specifying how network routing should be determined are not yet complete. Some specifications are in course of preparation (e.g. ATM Forum's *PNNI, private network–node interface*), but much work has still to be done. ATM network equipment manufacturers will therefore (at least for the meantime) have to design their own techniques. Ideally these will be simple and easy to manage, and lead to optimal network traffic loading at times of peak demand.

Once the route is setup for a particular connection, it is not usually altered during the duration of the call (i.e. the period of communication). Leaving the routing of the connection unaltered means that the transmission propagation time across the network between the two devices is not subject to any

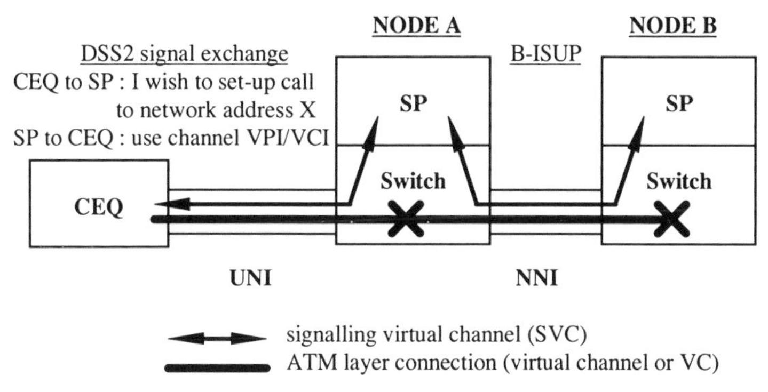

Figure 8.4 Setting up an ATM layer connection

unnecessary jitter (variability of delay). In addition, there is much reduced risk of cells which might otherwise have taken different paths from getting out of order. Finally, it is also much easier to determine and manage a network loading scheme, since nominal bandwidth allocations may be made to each of the connections which must statistically share a given physical transmission path. We discuss this point further later in the chapter.

8.6 ATM Network Topology State – the *Hello State* Machine

The ATM Forum is developing, as part of its *PNNI* (*private network–node interface* – based on the ATM *UNI v3.1)* specification, a sophisticated *source routing* control mechanism, in many ways similar to the techniques used in the *Internet.*

By keeping a record in its *topology database* of all information supplied to it about the topology of the network as a whole, an ATM network node always has a view of the entire private ATM network routing domain. The node is thus able to determine the route from itself to any *reachable* address.

The information about the topology and any changes made to it are conveyed as *PNNI topology state elements (PTSEs)*, including *topology state parameters*. These are conveyed between the nodes in the network by means of *PTSPs (PNNI topology state packets). Topology state parameters* are classified into two types:

- **Attributes** (these influence routing decisions – a security *attribute* of a particular node may cause the setup of a particular connection to be refused)
- **Metrics** (these are values which accumulate over the path of the connection as a whole to determine whether it is acceptable, e.g. the propagation delays of individual links in the connection are added as *metrics*)

When a new link or node is added to the network, then the directly affected nodes communicate with one another over the new link using the *hello* procedure. This is a standardized protocol enabling the two nodes to identify themselves to one another and work out the change in topology of the network as a whole. The new topology information is then *flooded* (i.e. broadcast) to the other nodes in their *peer group* (i.e. sub-network) or *advertised* to neighbouring *border nodes* of neighbouring peer groups by means of PTSPs. These inform the other nodes of any new addresses which are now *reachable* and also which exit route to take from the *peer group (PG)*.

The routing information is stored in the topology database as ‘address A *reachable* through entity B', where B is a known node within the peer group. Routing outside the *peer group*, as we shall see, is catered for by the *peer group leader (PGL)*.

A *peer group* as defined by ATM Forum's PNNI specification is a collection of nodes sharing the same *peer group ID (identifier)*. Since the peer group ID is defined during the initial configuration of the node by the human installer, a *peer group* is in effect a human operator-defined sub-network.

Within a *peer group* an *election* determines the *peer group leader (PGL)*. The *election* is an ongoing process which results in the node with the highest *leadership priority* taking over certain of the more important network routing (inter-sub-network) tasks on behalf of the peer group as a whole. The peer group leader is automatically promoted to the next layer in the routing hierarchy. However, should another node achieve a higher leadership priority (as a result of some topology or capacity change within the peer group) then it will take over the PGL function.

At the next layer of the hierarchy each *peer group* appears only as a single node – a *logical group node (LGN)*. It is represented in the topology

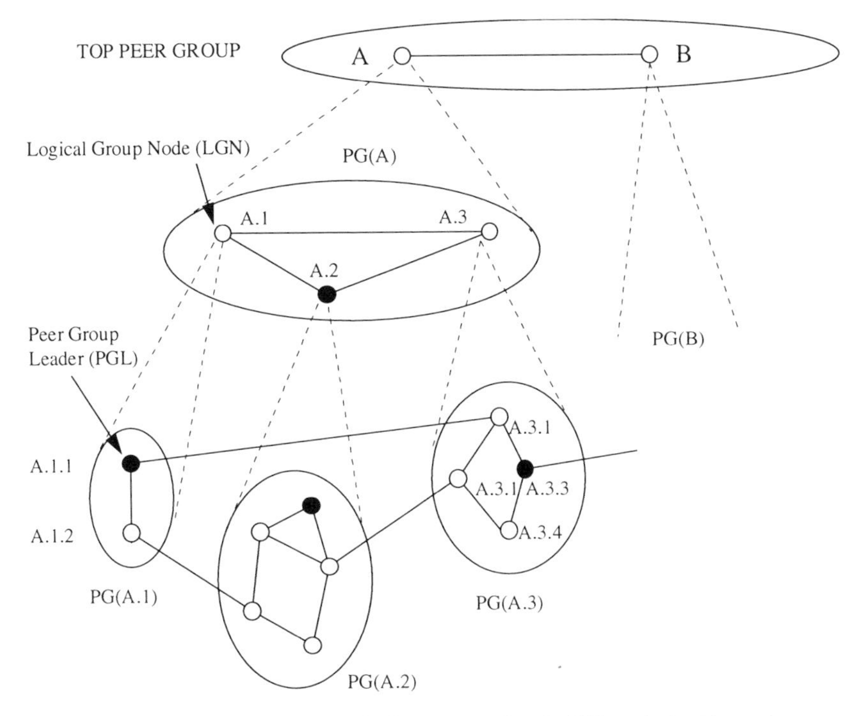

Figure 8.5 Peer groups (PG), logical group nodes (LGN) and the hierarchy of PNNI routing domains

management process at this level by the peer group leader. At the highest level in the hierarchy is the top peer group (Figure 8.5).

When neighbouring nodes running the *hello procedure* conclude that they belong to the same *peer group* then they synchronize their databases (recording the sub-network or *peer group* structure). They then *flood* (i.e. broadcast) this information to all other members of the peer group. In the case where the nodes do not belong to the same peer group, then they are *border nodes* in adjacent peer groups, and an *uplink* is said to exist from the border node to the *peer group leader* of the neighbouring *logical group node*. The communication between the border node and its partner's *peer group leader* is equivalent to the hello procedure – but this time between *logical group nodes* concluding that they are interconnected. The new topology information in this case is said to be *advertised* to other *peer group leaders* at the same hierarchical level. The *exterior reachable addresses (ERA*, i.e. those outside the peer group) are defined in a special routing table called a *designated transit list (DTL)* held by the peer group leader.

In contrast to *uplinks, horizontal links* are logical links between nodes in the same peer group.

Once the route to a given destination has been determined by a source node (either from its own database or from information provided by the peer group leader), normal UNI signalling procedures can be used to setup the connection. If necessary (e.g. due to current traffic conditions in the network causing a particular route to be unsuitable or overloaded) *crankback* and *alternative routing* may be invoked (in other words the first route choice is abandoned and a second path is attempted).

Before we leave the subject of topology databases and network routing, it is worth commenting on the form of labelling used to identify nodes in a PNNI network. The node names (or addresses) are similar in style to *Internet* addresses – lots of numbers, dots and letters. Thus the four nodes in peer group PG(A.3) of Figure 8.5 are called A.3.1, A.3.2, A.3.3 and A.3.4. This is the style of information held by the *topology database* and *designated transit list* (Address A *reachable* via B3.2, C4.4, D5.7, . . ., etc.).

8.7 Network Addressing Schemes Used in Support of the ATM Layer

ATM transport network addresses (*logical network addresses*) consists of three general parts, comprising nine separate fields as illustrated in Figure 8.6. The three main parts are the *AFI (authority and format identifier)*, the *IDI*

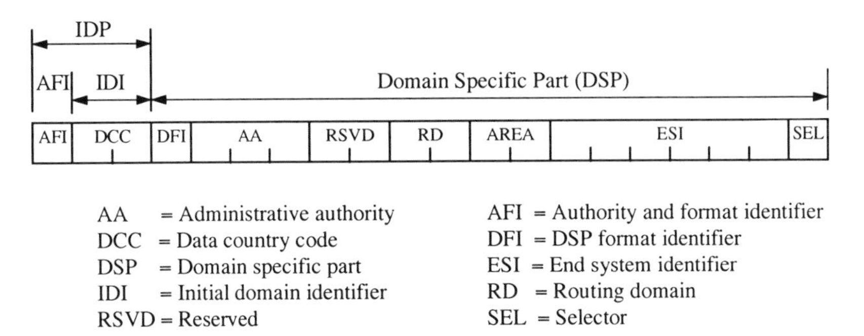

Figure 8.6 ATM addressing plans (conforming to OSI NSAP format)

(initial domain identifier) and the *DSP (domain specific part)*, where the AFI and IDP are known together as the *IDP (initial domain part)*.

The AFI allows for a number of different coding schemes to be used for the IDI, and determines which of these codings is in use. In the basic format, the IDI reflects the data country code as defined by ISO 3166 and ITU-T recommendation X.121. The remainder (the DSP) carries a number of fields identifying the administrative domain (i.e. network operator), the routing domain (sub-network), area code and end system number of the desired destination.

The *NSAP (network service access point)* format defined by ISO 8348 and ITU-T X.213 allows the AFI to be set to indicate one of three forms of the IDI. In the first, *DCC* form (AFI = 39), the IDI is the *data country code (DCC)* as defined by ISO 3166 (ITU-T X.121). In the second form (AFI = 47), the *ICD* or *international code designator* form, the IDI identifies international organizations as defined by British Standards. In the third form (AFI = 45), the IDI takes the form of a full E.164 international ISDN (telephone) number, and the separate fields DFI and AA do not appear.

The DFI defines the format (and therefore the meaning) of the DSP (domain specific part). The *administrative authority (AA)* identifies the organization which administer the addresses in this field (usually the public or private network operator). The *routing domain (RD)* together with the *area* field indicate the destination sub-network and area. The *ESI (end system identifier)* indicates the end device. The *selector* field is an extension of the address; it is not used for routing within the network, but may be required to indicate to the end system which local device (e.g. telephone extension) is to be connected or which mode of communication should be expected (e.g. telephone, fax, etc.).

8.8 Signalling Functions and Procedure in Support of the ATM Layer

Connections across an ATM transport network (user plane), as we have seen, are established and controlled by signalling transactions which are conveyed by the *control plane*. Higher layer *signalling protocols* carried from the customer equipment (CEQ) to the network by the control plane request the setup of an ATM layer connection on the user plane. The control plane signalling also manages the connection during communication and arranges *clearing* of the connection after communication has ceased. The signalling messages, as previously discussed are defined in DSS2 (digital subscriber signalling system 2 – used across the UNI) and B-ISUP (broadband integrated services user part – used across the NNI).

During connection setup a negotiation between CEQ and network takes place in which the CEQ first identifies the end device to which it wishes to be connected, the type of connection it requires, the quality of connection required, the peak cell-rate required (i.e. bandwidth) and any supplementary services needed. The CEQ *negotiates* with the network according to the availability of bandwidth to the destination, and, by means of the signalling channel (SVC), is allocated the virtual channel (i.e. VPI/VCI combination) which is to be used for the connection.

Signalling messages in DSS2 (as defined by ATM UNI v3.1 and ITU-T Q.2931) can be grouped into five main types as listed in Table 8.3.

Each signalling message contains a number of *information elements (IE)*. These contain information such as the network address of the calling port and intended destination port, the peak cell rate and quality of service required and so on. In addition, all signalling messages related to a particular attempt to establish a connection (a *call attempt*) are numbered with the same *call reference.*

Figure 8.7 illustrates in broad terms the signalling sequence for setting up a point-to-point switched virtual circuit (SVC) connection. The calling CEQ (the *A-party*) sends a SETUP message to the network. The network considers the routing to the destination and signals via further network nodes using B-ISUP. Meanwhile, the A-party is informed of the state of continuing call setup by the CALL PROCEEDING message. This prevents any timeouts (or other impatience on behalf of the A-party device). Finally, when the final network node has been reached, the destination node signals to the destination device (the *B-party*) with a SETUP message in which a VPI/VCI value is given. This is the equivalent of ringing a telephone bell. The VPI/VCI value tells the B-party device on which virtual channel to look.

By accepting the connection (or answering the call), a CONNECT message is generated by the B-party. The network immediately responds to it with

Table 8.3 Signalling message types at the ATM UNI (Digital subscriber signalling system 2–DSS2)

Signalling message type	Signalling message name
Call establishment messages	SETUP CALL PROCEEDING CONNECT ACKNOWLEDGE
Call clearing messages	RELEASE RELEASE COMPLETE
Status messages	STATUS ENQUIRY STATUS (response)
Global call reference related messages	RESTART RESTART ACKNOWLEDGE STATUS
Point-to-multipoint connection controls	ADD PARTY ADD PARTY ACKNOWLEDGE ADD PARTY REJECT DROP PARTY DROP PARTY ACKNOWLEDGE

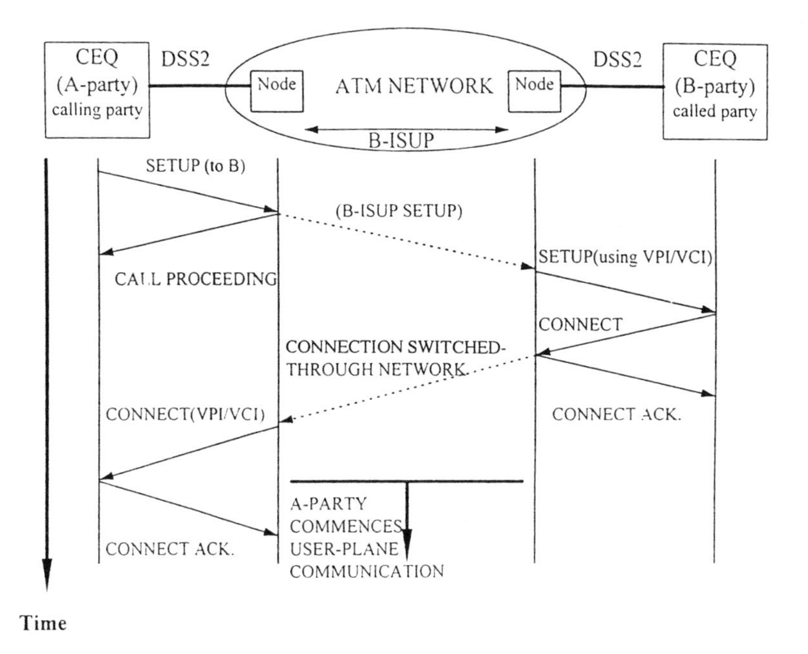

Figure 8.7 Signalling interchange necessary for the setup of an ATM layer (user plane) connection

CONNECT ACKNOWLEDGE (CONNECT ACK. for short). This confirms receipt of the message and in effects asks the B-device to hold on while the connection is setup to the A-party. The message cascades back through the network until reaching the originating network node, which informs the A-party by allocating a VPI/VCI value for the communication. The A-party device confirms with a CONNECT ACKNOWLEDGE, and communication can then take place at the ATM layer of the *user plane* directly, between the end devices (A and B parties). The *user plane* AAL entities within these two devices can then establish an AAL service and convey higher layer protocols between one another.

A similar signalling sequence to that depicted in Figure 8.7 takes place when communication is over, in order to clear the connection. In this case, RELEASE messages replace the SETUP messages, and RELEASE COMPLETE messages replace the CONNECT messages. There is no equivalent of the CALL PROCEEDING and CONNECT ACKNOWLEDGE messages.

ATM network signalling also allows the establishment of point-to-multipoint connections. Such connections are useful means of broadcasting the same information to several end-users. Examples of services which could benefit from point-to-point connections are broadcast TV. The procedure for setting up multipoint connections is similar to that for point-to-point connections, except that after the connection of the first B-party, further called parties are added to the connection by using the point-to-multipoint connection control messages ADD PARTY and ADD PARTY ACKNOWLEDGE. These take the place of SETUP and CONNECT messages respectively.

The carriage of DSS-2 and B-ISUP messages (as we already briefly discovered in Figure 7.4) is dependent upon the SSCF (service specific coordination functions), SSCOP (service specific connection-oriented protocol) and AAL5. These are discussed more thoroughly in chapter 9.

8.9 Connection Quality Control at the ATM Layer

We have already discussed the negotiation which takes place at the time a calling CEQ *attempts* to set up a connection across an ATM transport network. This is one of the most important quality control features of ATM. The need for it arises from the statistical multiplexing technique (cell relay) upon which ATM is based.

Figure 8.8 illustrates the variation in cell delay which arises due to the statistical nature of the cells arising from the various traffic sources. While over a period of time (longer than that illustrated) the net arrival rate of cells is planned to be lower than the cell carrying capacity of the trunk (or network as a whole), there will arise periods (as shown) when an accumulation of cells occurs and individual cell transit delays increase momentarily. In order to

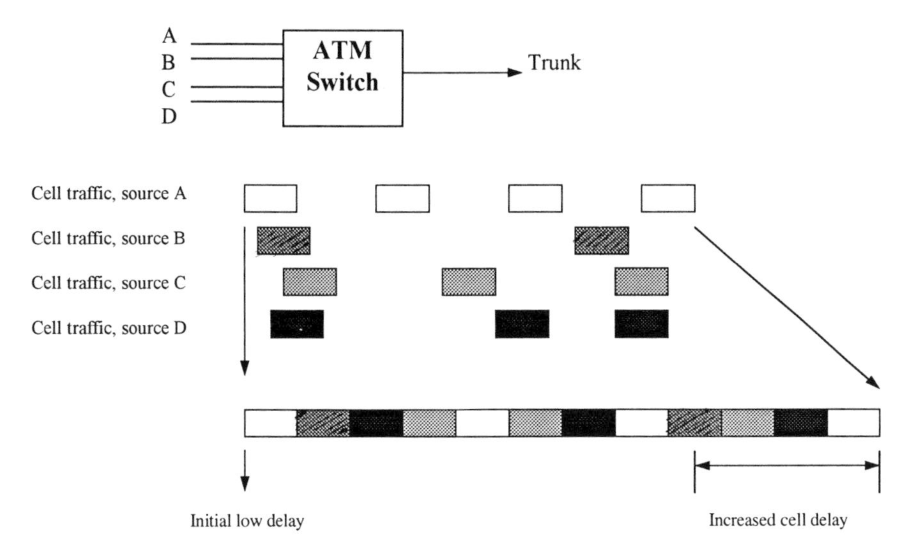

Figure 8.8 Cell delay variation in an ATM network

assure a reasonable quality of service for the individual connections carried by an ATM network it is necessary to operate traffic and congestion control measures. The measures to be used are laid out in ITU-T recommendation I.371.

Traffic and congestion control in ATM networks is achieved by two measures:

1. A negotiation process (*connection admission control, CAC)* is carried out at each call setup, in which the network allocates appropriate capacity to cater for the required *peak cell rate* and connection quality of service or else rejects the new call.
2. In addition, a network congestion monitoring and relief process (*network parameter control, NPC* and *usage parameter control, UPC*) is carried out during the active period of the connection. This discards less important cells at times of congestion and implements flow control procedures in order to relieve congestion.

Each time a connection is requested from the ATM layer (at the ATM SAP, *service access point*), the requesting entity (be it the AAL or some other user) is required by the *connection admission control (CAC)* procedure to declare its connection type needs in terms of the following parameters:

- *Peak cell rate (PCR)* required
- *Sustainable cell rate (SCR)* required (i.e. the minimum persistent transport required)

- *Burst cell rate* (i.e. the maximum cell rate)
- *Quality of service (QOS),* defined in terms of the parameter's cell delay, sensitivity *to cell delay variation tolerance (CDVT)* and *cell loss ratio (CLR)*

The parameters are carried in the call setup message and form part of a *traffic contract* which the network commits to at the time of connection establishment. It is a commitment to the end-user or device that the network will meet the requested quality of service, provided the user complies with the conditions he has specified. Such *traffic contracts* are negotiated between end user devices and the network at the UNI, and where connections traverse multiple networks, also at each *inter-network interface (INI).*

Currently there are no standards which guide the network equipment designer regarding how many traffic contracts the network may enter into (there is, for example, no guidance to say by how many percent the network capacity may be 'overbooked'). The network equipment manufacturers, however, are all developing their own algorithms to determine the *equivalent capacity* of individual connections in order that the statistical likelihood of cell overloads is kept within manageable bounds, while simultaneously allowing 'overbooking' in order that there is an efficiency gain achieved by the statistical multiplexing. It would, after all, make no sense to allocate the network resources (i.e. trunk bandwidths) according to the peak cell rates required for the individual connections. This would ensure no cell collision and loss, but equally no statistical multiplexing gain from use of the free slots between bursts of peak cell rate sending. The process of bandwidth allocation is termed *network resource management (NRM)* by the ATM standards.

Once the traffic contract is established and the connection is set up, the *usage parameter control (UPC)* and *network parameter control (NPC)* procedures take up the process of monitoring network performance and service delivery according to the contract. Their main purpose is to protect the network resources and network users from quality of service degradations arising from unintentional or malicious violations of negotiated traffic contracts, and take appropriate action. These control procedures are based on the *generic cell rate algorithm (GCRA)* which is used to determine whether a given cell stream is within its contract or not.

On detecting a violation the UPC or NPC may elect to carry the extra cells anyway, may reschedule the cells, may discard them, or may *tag* them, by overwriting the *cell loss priority (CLP)* bit, resetting the value from '0' to '1' (a process known as *tagging*) and therefore increasing the probability of cell discarding should congestion be encountered.

When congestion does arise in the network, measured in terms of excess *delay*, excess *cell delay variation* or some other measure, then the ATM layer responds with one of four actions:

- *traffic shaping,*
- *fast resource management,*
- *selective cell discard*
- *explicit forward congestion indication (EFCI).*

Traffic shaping alters the traffic characteristics of a given stream of cells by reducing the peak cell rate, limiting the burst length of peak cell rate sending, reducing cell delay variation by regulating the submission of cells or some other form of queueing adjustment mechanism.

Fast resource management is the term given to the reallocation of network-wide resources (i.e. bandwidth) to meet the instantaneous cell traffic demand. The reallocation of resources should occur within the round trip propagation delay of the ATM connection. In theory it could then adjust the use of network resources to cope with short duration bursts. It is a fine goal indeed, but is not yet well defined and is likely to be difficult to realize, particularly in large and complicated networks.

The *explicit forward congestion indicator (EFCI)* is a notification message to the destination CEQ that the network is congested. The idea is that the destination CEQ should then respond (if capable) by implementing protocols with the sending device to adapt (i.e. lower) the cell rate until the EFCI signal is lifted. The signal arises from a similar mechanism which is implemented in the *frame relay* protocol. It is a very effective way of avoiding and alleviating congestion for data communications devices.

Selective cell discard is the final resort. If all else fails, then the only means to alleviate congestion is to live with the inevitable consequence – loss of information. The *cell loss priority (CLP)* bit in the cell header enables *priority control (PC)* to be applied to the discarding of cells. This ensures that higher priority cells (CLP = 0) are more immune to congestion than lower priority cells (CLP = 1). In very serious congestion, of course, even high priority cells will be discarded.

8.10 ATM Layer Management

ATM network management we discuss in more detail in chapter 10. However, in order to complete the subject of the ATM layer in this chapter, we should note that the layer management protocol (using the terminology of the protocol reference model) is called ATM management (ATMM). It is

Table 8.4 Cabling alternatives for connecting user devices at the ATM UNI

Optical cable	Coaxial cable	Twisted pair cable	
Physical characteristics	Monomode fibre pair conforming to ITU-T G.957, operating on 1310 nm laser signal wavelength	75 Ω coaxial cable pair, one cable for transmit, one for receive	Unshielded twisted pair category 3 (UTP3) for speeds up to 52 Mbit/s. Unshielded twisted pair category 5 (100)UTP5) or 150 Ω shielded twisted pair (STP) for speeds of 100 Mbit/s and 155 Mbit/s. One pair each for transmit and receive
Maximum connection length	Typicall 15 km	Recommended less than 100 metres	Approx 100 metres with type 1 or type 2 cable
Media Interface Connector (MIC–the plug, jack or socket)	BFOC/2.5 connector plugs according to (ISO/IEC 86B)	BNC connector	RJ-45 (8-pole) plug and socket (ISO/IEC 88uu). Contacts 1 and 2 are transmit + & − from ATM user device, contacts 7 and 8 are transmit + & − from ATM network. Alternatively the 9-pin shielded D-plug and socket may be used. In this case the user equipment shall transmit on pins 5 (+) and 9 (−), and the ATM network shall transmit on pins 1 (17) and 6 (609)
Optical line speeds	155.52 Mbit/s	34,348 Mbit/s	12.96 Mbit/s 25.92 Mbit/s 51.84 Mbit/s 155.52 Mbit/s
Line code	NRZ (non return to zero) code light emission = binary '1'	CMI (coded mark inversion) code conforming to ITU-T G.703	For speeds up to 52 Mbit/s, CAP-16 code. For 155 Mbit/s, 8B/10B code

conducted by the ATMM entity. Its functions are defined in ITU-T recommendation I.150.

8.11 Physical Layer Specification (B-ISDN UNI and NNI)

ITU-T recommendation I.432 defines the physical layer interface to be used at the UNI (reference points S_B and T_B). Optical fibre is defined to be the preferred medium but specifications are also provided for an electrical, coaxial cable interface and a twisted pair copper cable interface (defined by ATM Forum).

There is no current formal specification for the physical layer interface at NNI, but this will be strongly based on, if not identical to, the UNI physical layer interface.

The characteristics of the various optional physical layer interfaces are listed briefly in the Table 8.4.

Which device should carry a socket or a plug is defined in terms of two *interface points*, defined as I_a and I_b by the standards. These are imaginary connection points, respectively on the user and on the network side of the S_B and T_B reference points, as illustrated in Figure 8.9.

8.12 The Transmission Convergence Sublayer (Physical Layer)

In addition to the mechanical and electrical aspects of the physical interface (the *physical medium (PM)* sublayer), ITU-T recommendation I.432 also defines a procedural layer between the user end device (B-TE) and the network termination device (B-NT1), called the *transmission convergence sublayer (TC)*.

The transmission convergence sublayer defines a number of rules for the

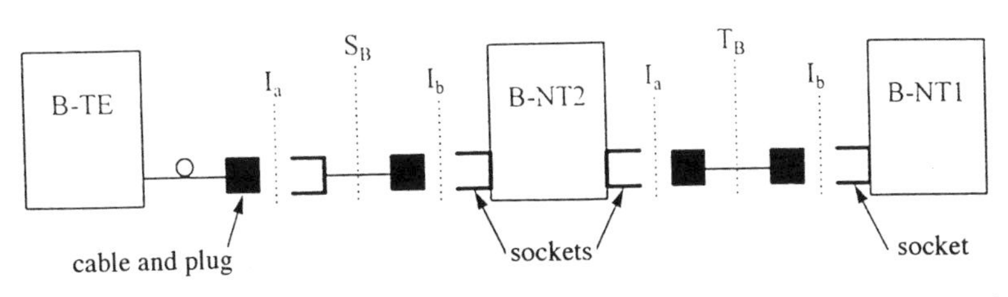

Figure 8.9 Interface points at the ATM UNI and their standardized plug/socket interface

transmission of cells across the physical medium which ensure their correct transmission and proper interpretation. Functions provided by the transmission convergence sublayer include, for example, cell delineation – the identification of cell boundaries and cell scrambling/descrambling. Cell scrambling randomizes the cell payload, thereby minimizing the chance of continuous non-varying bit patterns which could, for example, intentionally or unintentionally emulate network status or control signals, or otherwise disturb the network.

The transmission convergence sublayer is realized by extra cells added into the ATM layer cell stream (i.e. into the user's information-containing cells). After each 26 *ATM layer cells*, a *physical layer (PL) cell* is added, so that every 27th cell is a PL-cell. This provides an adjustment from the 155520 kbit/s bitrate at the T_B-interface so that the maximum bit rate available for user information transfer is 149760 kbit/s (corresponding to the SDH C4-container).

Where the NT1 provides an interface between an optically connected B-TE operating in a *cell stream mode* and an SDH network, the transmission convergence sublayer reduces the capacity of the cell stream to 149760 kbit/s and carries out certain framing, delineation and error correction tasks. Of course on the network side of the B-NT1 equivalent tasks of framing and delineation are adequately catered for by the *message overhead (MOH)* of SDH (as we discuss in appendix 1), so the full 149 760 kbit/s of user information can be passed directly into the SDH frame.

'Idle' PL-cells are additionally inserted instead of ATM-cells when there are no ATM cells to be sent.

At least every 513th cell is a *PL-OAM cell* (operation and maintenance cell). OAM cells are classified as F1 (regenerator level), F2(digital section level) and F3(transmission path level) cells. These carry out analogous functions to the message overhead of SDH. The coding of the PL-OAM cells includes a number of different fields and functions designed to detect errors, monitor system performance and send network failure alarms. Table 8.5 summarizes the field functions of PL-OAM cells. When no failures are present, then only the fields CEC, MBS, NMB, NIC and PSN need be used.

Physical layer cells (PL-cells and PL-OAM cells) always have the CLP (cell loss priority) bit in the ATM cell header set CLP = 1, with the VPI and VCI both set at all zeros. Idle cells are marked with payload type, PTI = 000. PL-OAM cells are marked with PTI = 100.

8.13 Carriage of ATM Direct over SDH Transmission Lines

As well as using a cell-oriented transmission medium (e.g. optical fibre connected directly between ATM devices (CEQ/ATM node or ATM node/ATM

Table 8.5 OAM cell fields at the transmission convergence sublayer

Acronym	Field or signal name	Purpose
CEC	Cell Error Control	A cyclic redundancy code is used to detect errors in the cell payload
EDC	Error Detection Code	A detection mechanism for block errors
LOC	Loss of Cell delineation	A network status which causes generation of the remote defect indication (RDI) signal
LOM	Loss of OAM cell	A network status which causes generation of the remote defect indication (RDI) signal
MBS	Monitoring Block Size	The block size based upon which block error detection is performed. The smaller the MBS the greater the error accuracy but the greater the overhead and the lower the efficiency
NMB-EB	Number of Monitored Blocks – Far End Block errors	Indicates the number of blocks between this OAM cell and the last, for which EDC s are being sent in the following octets
NIC	Number of Included Cells	Indicates the number of cells between this OAM cell and the last
PSN	PL-OAM Sequence Number	Enables identification of cells, checking of receipt and could enable regeneration of lost messages
R	Reserved Field	A currently unused field containing the idel pattern '01101010' reserved for future use
S-AIS	Section Alarm Indication Signal	A signal used to alert equipment in the direction of transmission that an alarm has been detected on the transmission section
S-RDI (formerly S-FERF)	Section Remote Defect Indication (formerly Section Far End Received Failure)	A signal returned to the sending end when an LOC or LOM is detected, equivalent to the AIS sent to the receiving end of a transmission section
TP-AIS	Transmission Path Alarm Indication Signal	A signal used to alert equipment in the direction of transmission that an alarm has been detected on the transmission path
S-FEBE	Section Far End Block Error	A signal reporting parity violations in the block as detected at the distant end of the transmission section
TP-FEBE	Transmission Path Far End Block Error	A signal reporting parity violations in the block as detected at the distant end of the transmission path
TP-RDI (formerly TP-FERF)	Transmission Path Remote Defect Indication (formerly Transmission Path Far End Received Failure)	A signal returned to the sending end when an LOC or LOM is detected, equivalent to the AIS sent to the receiving end of a transmission path

node)), ITU-T recommendation I.432 also defines an SDH (synchronous digital hierarchy) compatible interface. In this case the transmission convergence sublayer (TC) is replaced by the normal path and message overheads provided within SDH.

8.14 ATM Forum's *Utopia* specification

The *Utopia* specification (*universal test and operations physical interface for ATM*) was laid down by ATM Forum in March 1994. Unlike ITU-T, ATM Forum has in this specification attempted to define a common interface between the physical and ATM layers used in ATM equipment. The intention is to allow greater interchangeability of equipment, particularly the use of common ATM switching chips.

9
The ATM Adaptation Layer (AAL)

The ATM Adaptation Layer (AAL) is perhaps the most important functional part of ATM, since it provides for the conversion of information into a format suitable for carriage across an ATM network. In this chapter we discuss the various types of AAL (e.g. AAL1, AAL2, AAL3/4, AAL5) which have so far been defined, describing the function of each and when it should be used. Later in the chapter we describe also the various service specific convergence sublayer (SSCS) functions which provide for mapping of different types of services and communication protocols (e.g. frame relay, SMDS) into a format suitable for carriage by an ATM network.

9.1 Function of the *ATM Adaptation Layer (AAL)*

The *ATM adaptation layer (AAL)* provides for the conversion of information into a form suitable for carriage by an ATM transport network. AAL functionality is provided on either side of an ATM transport network (as shown in Figure 9.1). It provides (at the sending end) for the conversion of a telephone, data or other communication signal into the ATM cell format. At the receiving end, the AAL reverses the conversion, turning the cells back into the original signal.

Figure 9.2 illustrates the same thing, but in the manner of the protocol layer model. In this diagram you must understand that the in-trade jargon *higher layer information* is used to mean the user information (e.g. telephone speech signal, video signal, frame relay encoded data signal or other information) to be carried across the network. The higher layer information is passed to the ATM adaptation layer (AAL) which converts it into the form in which the

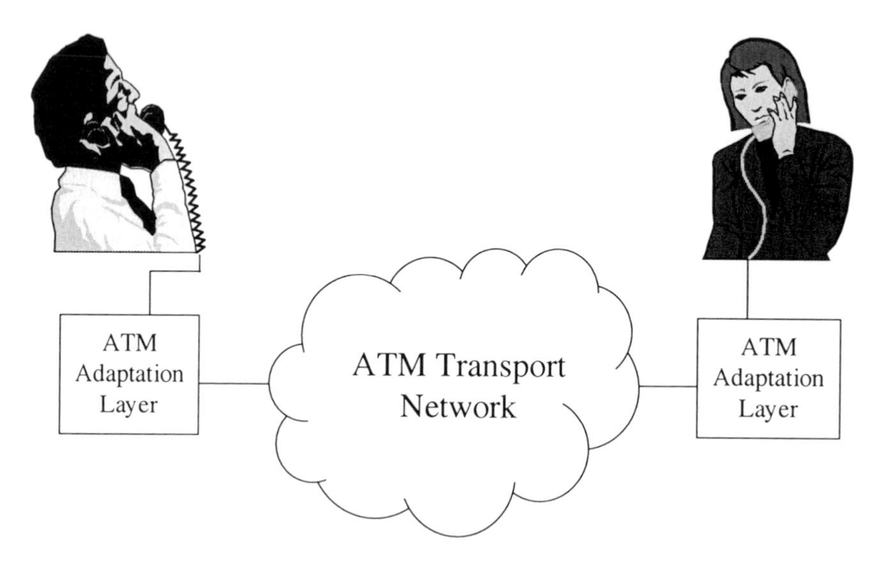

Figure 9.1 Function of the ATM Adaptation Layer (AAL)

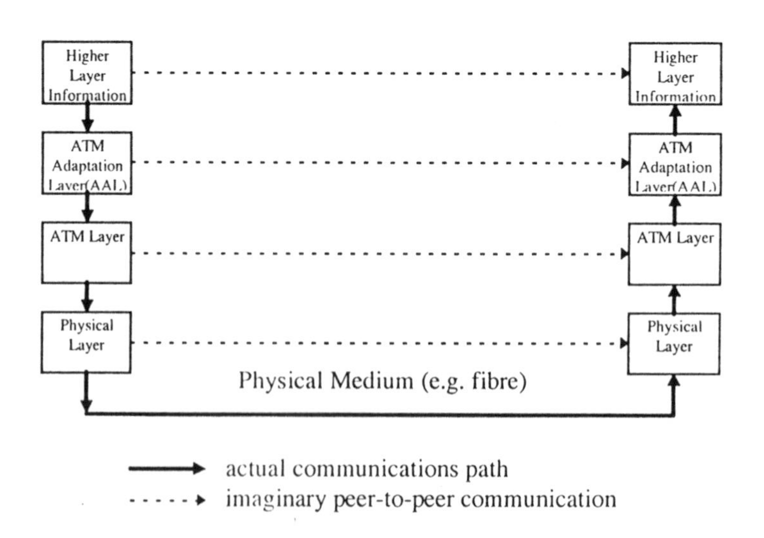

Figure 9.2 Protocol Reference Model representation of the ATM Adaptation Layer (AAL)

ATM transport network (i.e. ATM and physical layers) can cope with it. As we discussed in chapter 7, it is the AAL-PDU (AAL protocol data unit, known by the ATM layer as the ATM-SDU, ATM layer service data unit) which is suitable for carriage by the ATM layer. The ATM layer converts the ATM-SDU into the ATM-PDU (ATM protocol data unit – equivalent to the PL-SDU (physical layer SDU)). Finally the physical layer sees to the physical transmission across the line. The receiving protocol stack is traversed from bottom to top, reconverting the signal into the original higher layer information.

The ATM adaptation layer serves to perform conversion of the higher layer information into cells and for the reconstruction of the original signal. The AAL has to make sure that all the cells are received and are in the right order and that the error rate in the received cells is within acceptable bounds, given the nature of the user information being carried. For information signals which are sensitive to cell delay variation (CDV), the AAL provides a buffering and clocking functionality in order to make sure that the signal sent to the destination device is free of CDV. CDV in the case of a telephone signal, for example, manifests itself undesirably as *jitter*.

9.2 The Sublayers of AAL

The functionality of the AAL is divided between two sublayers (Table 9.1). These are the *convergence sublayer (CS)* and the *segmentation and reassembly (SAR) sublayer*. It is the convergence sublayer which accepts the AAL-SDU (i.e. the higher layer user information) and processes it into cells. Since the conversion may comprise a service specific part and a part which is common to a number of similar communication techniques, the conversion sublayer is also further subdivided into the *service specific convergence sublayer (SSCS)* and the *common part convergence sublayer (CPCS)*. An SSCS exists, for example, for frame relay (FR-SSCS, allowing frame relay signals to be

Table 9.1 Sublayers and service access points of the ATM Adaptation Layer (AAL)

AAL Service Access Point (SAP)		
ATM Adaptation Layer (AAL)	Convergence Sublayer (CS)	Service specific convergence sublayer (SSCS) Common part convergence sublayer (CPCS)
	Segmentation and Reassembly (SAR) Sublayer	

converted to an ATM transportable format). It would share a CPCS with X.25 had this been defined.

The *segmentation and reassembly (SAR)* sublayer of the AAL provides for the *sequence numbering* of the cells passed to it by the CS and their passing to the ATM layer as an SAR-PDU (equivalent to the ATM-SDU) at the ATM layer *service access point (SAP)*.

The various different common part convergence sublayers (CPCS), together with their respective SARs, are classified into four different AAL types as shown in Table 9.2. The distinguishing characteristics of the various types of AAL are also indicated in the table.

You may wonder what happened to AAL types 3 and 4, and why there is an AAL type 5, when there are only four classes of AAL. The reason lies in the history of the evolution of the standards. Originally, the standards had foreseen four different connection type classes (A to D) and had numbered the AAL types 1 to 4 respectively to correspond with these connection classes. Most work was then applied to AAL types 1, 3 and 4 since these correspond to communications services where there is potential for ATM to substitute existing networks. During the course of development AAL3 and AAL4 converged, and it was decided to merge the two types in order to simplify the technical realization. AAL3/4 resulted.

Table 9.2 Service classification of the ATM Adaptation Layer (AAL)

	AAL classes			
Transmission characteristic	Class A	Class B	Class C	Class D
AAL type	AAL Type 1 (AAL1)	AAL Type 2 (AAL2)	AAL Type 3/4 (AAL3/4) AAL Type 5 (AAL5)	AAL Type 3/4 (AAL3/4), AAL Type 5 (AAL5)
Timing relation between source and destination	Required	Required	Not required	Not required
Bit rate	Constant	Variable	Variable	Variable
Connection mode	Connection-orientated	Connection-oriented	Connection-oriented	Connectionless
Example communication services within this class	Circuit-switched or leaseline-like connections (e.g. telephone, E1, T1, n × 64 kbit/s etc.)	Packet audio or video signals	Frame relay, X.25 etc.	IP (Internet), SMDS etc.

AAL5 performs very similar functionality to AAL3/4 but is a much simpler protocol, not requiring to add much protocol control information (PCI). As a result it uses less of the line bitrate capacity for overhead information, and thus allows proportionately more user information to be carried. This is the root of the original name for AAL5, the *simple efficient adaptation layer (SEAL)*.

In comparison with AAL5, AAL3/4 provides for a more sophisticated form of multiplexing different higher layer information within the AAL for carriage by the ATM layer over the same VPI/VCI connection to a common end device. The price is the much higher protocol overhead which is needed. This may be important for certain types of services (e.g. SMDS), but for the development of protocols for the carriage of simpler services (e.g. frame relay), AAL5 is generally preferred, due to its higher efficiency.

During the setup of a connection, the DSS2 CONNECT message contains an *AAL information element (AAL-IE)* in which the calling end notifies the called end of the type of AAL which will be used. The format and coding of the AAL-IE are shown in Table 9.3.

Table 9.3 The ATM Adaptation Layer parameter information element and coding

Bits								
8	7	6	5	4	3	2	1	Octets
0	1	0	1	1	0	0	0	1
Information Element Identifier								
1	Coding standard			IE Instruction field				2
Length of ATM parameters								3
Length (continued)								4
AAL type								5
Further information								6, 7, ...

Field	Optional code settings	
Coding standard	Always set as 00	
IE Instruction field	Always set as 00000	
AAL type	0000 0001	AAL type 1
	0000 0010	AAL type 2
	0000 0011	AAL type 3/4
	0000 0101	AAL type 5
	0001 0000	User-defined AAL

9.3 AAL Type 1 (AAL1)

AAL Type 1 provides for the carriage of constant bit rate, delay sensitive connections across an ATM network. Using AAL1, circuit-switched and leaseline connections may be carried across ATM networks, using the ATM *constant bit rate (CBR)* service. The type and bitrate of the CBR service required are defined in octets 6 and 7 of the AAL-IE (AAL information element) at the time of connection setup. Acceptable values are given in Table 9.4.

Table 9.4 The various AAL1 subtypes

Octet	Field	Bits								Meaning
6	AAL subtype	0	0	0	0	0	0	0	0	Null
		0	0	0	0	0	0	0	1	Voice band transport
		0	0	0	0	0	0	1	0	Circuit emulation
		0	0	0	0	0	1	0	0	High quality audio
		0	0	0	0	0	1	0	1	Video
7	CBR rate	0	0	0	0	0	0	0	1	64 kbit/s
		0	0	0	0	0	1	0	0	(T1) 1 544 kbit/s
		0	0	0	0	0	1	0	1	(T2) 6 312 kbit/s
		0	0	0	0	0	1	1	0	32 064 kbit/s
		0	0	0	0	0	1	1	1	(T3) 44 736 kbit/s
		0	0	0	0	1	0	0	0	97 728 kbit/s
		0	0	0	1	0	0	0	0	(E1) 2 048 kbit/s
		0	0	0	1	0	0	0	1	(E2) 8 448 kbit/s
		0	0	0	1	0	0	1	0	(E3) 34 268 kbit/s
		0	0	0	1	0	0	1	1	(E4) 139 264 kbit/s
		0	1	0	0	0	0	0	0	n × 64 kbit/s
		0	1	0	0	0	0	0	1	n × 8 kbit/s

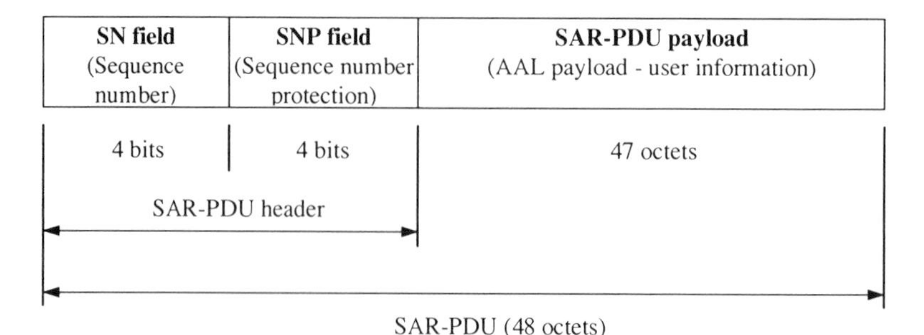

Figure 9.3 SAR-PDU format of AAL type 1

The 48-octet payload of the ATM layer cell is subdivided by AAL1 into a 47-octet AAL payload (correctly the SAR-PDU payload) and a one-octet AAL1 header (Figure 9.3).

The first bit of the *sequence number (SN)* field is the *CSI (convergence sublayer indication) bit.* This is used to indicate the use of convergence sublayer protocol control information. In other words, some of the 47 octet SAR-PDU payload has been used for CS PCI in order to provide for some sort of convergence sublayer function.

The CSI bit is followed in the SN field by a 3-bit *sequence count field.* This is used to number consecutive cells of a particular constant bit rate source information stream. The sequence count is used by the receiving SAR to check that all cells have been received and that cells are in the correct order. (Cells could, for example, be lost or get out of order at a time of network trunk failure, if an automatic connection re-route took place during the call. In addition, technical problems in the buffers of intermediate ATM layer switches could cause jumbling of the cell order.)

The *sequence number protection (SNP)* field provides error detection and correction capabilities for the SAR-PDU header. This ensures correct interpretation of the SN field information even if it is corrupted by transmission line errors. The SNP field comprises a 3-bit CRC (cyclic redundancy check) code and an even parity bit.

The SN and SNP fields together make up the protocol control information (PCI) of the segmentation and reassembly (SAR) sublayer.

For technically-minded readers, the content of the 3-bit CRC field is the remainder of the modulo 2 division by the generator polynomial $X^3 + X + 1$ of the product X^3 multiplied by the content of the SN field. The even parity bit is set to '1' or '0' to make sure that an even number of 1's appear in the 8 bits of the SN and SNP fields.

The *common part convergence sublayer (CPCS)* of AAL1 provides for the acceptance of a constant bitrate infomation stream and its conversion into 47-octet cells. (The number of cells generated per second depends upon the required CBR service bitrate.) At the receiving end, the CPCS of AAL1 processes the *sequence count number* of the SAR header to check for cell loss or misinsertion of cells. The other primary function of the CPCS for constant bit rate circuit transport is the handling of cell delay variation (CDV), in order that the bitrate of the received signal is as constant as that sent. This is achieved using a receive buffer to remove the variation, together with a technique called *source clock frequency recovery* to ensure correct bitrate of the arriving signal at the destination user equipment.

The recommended size of the buffer used to eliminate *cell delay variation (CDV)* is dependent upon the type of source user information. The advantage of a relatively large buffer is that the received bit rate can be kept very stable. The disadvantage of a large buffer, however, is the greater mean transmission delay. Too much delay in itself may be just as much of a problem as cell delay variation. In the case of telephone speech, for example, delay much more than 50 ms disturbs the course of conversation, and may lead to additional problems, including echo.

In the event that cells arrive at the destination slightly too fast for a while then the buffer will gradually fill up until the cell arrival rate reduces again. Should it not, then the buffer will overflow, and the receiving CS (convergence sublayer) will have to react as if cells have been lost. Should the cell arrival rate be too low, then the buffer may underflow (i.e. run out of bits). In this instance, the receiving CS should generate an appropriate number of dummy bits in order to maintain the bitrate.

When cells arrive at the receiving CS out of order or foreign cells stray in by mistake from other connections (i.e. are misinserted – perhaps because of corruption of the header information), they are detected by means of the *sequence count field.* Misinserted cells are simply discarded. Lost cells, resulting from sequence count errors or buffer underflow, are replaced by dummy bits in order that the received bit rate is correct.

You may think that it is rather crude to discard misinserted cells and replace lost cells with dummy information, but for *constant bit rate (CBR)* service, this is a more appropriate action than trying to recover the correct order of cells or resending lost cells. By buffering to correct insertion errors or resending, a considerable extra delay is inflicted on the signal, and maybe also a high *cell delay variation (CDV).* For telephone speech, the dummy information which replaces the lost or misinserted cells may result in a faint 'click' but this is likely to be barely noticeable to the listener, provided the loss rate is not high. For a point-to-point video connection, a spot may appear on the received picture momentarily. For a point-to-point data line using a CBR connection, the higher layer protocol will detect and correct the error anyway.

Source clock frequency recovery ensures that the bitrate of the signal received from the source equipment is conveyed exactly to the receiving device. There are two alternative methods for *source clock frequency recovery*:

- The *synchronous residual time stamp (SRTS)* method
- The *adaptive clock* method

In the *SRTS* method both sending and receiving devices have access to a common reference clock, typically the clock provided by the network to which they are both connected. The sending device compares its own clock with that of the reference over a fixed number of SAR-PDU payloads. Having determined the relative frequency difference a value, termed the *residual time stamp (RTS)*, is calculated and transmitted to the receiver. By comparing and adjusting its own clock frequency against that of the network reference clock, the receiving device is able exactly to match the sending frequency. The RTS value is sent as 4 bits of CS protocol control information in the first octet of the SAR-PDU payload. It may only be sent when the sequence count is odd (i.e. value 1, 3, 5 or 7) and when the CSI (convergence sublayer indication bit) is set to '1'.

In the *adaptive clock method* the fill level of the receiving AAL buffer is used to control the frequency of the local clock. Thus as the buffer fills up the local frequency is increased, so as to compensate. The adaptive clock does not require CS protocol information to be sent, and in addition may be used where the connection between sender and receiver traverses several ATM networks, the clocks of which are not precisely synchronized with one another.

When AAL1 is used to support certain types of PDH (plesiochronous digital hierarchy – see appendix 1) multiplex structured bitrates, it is helpful to indicate by CS protocol information the start of each structured block (i.e. each multiplex frame). This is done by the use of a *pointer* carried in the first octet of the SAR-PDU payload when CSI = 1 and the sequence count is an even value (i.e. 0, 2, 4 or 6). Carriage of such bitrates using AAL1 is termed *structured data transfer (SDT)*.

The length of the AAL-SDU for AAL1 is set, depending on signal input type, to either 1 octet (8 bits) or 1 bit. Thus, for a standard 64 kbit/s telephone voice channel, which is composed of 8000 Hz times 8 bits, each 8-bit voice signal *codeword* can be passed directly to AAL1 for carriage over ATM.

9.4 AAL Type 2 (AAL2)

AAL2 is the least well developed of the common AAL types. This reflects the relatively few existing services for packet carriage of video and class B data signals. In principle, AAL2 could be similar to AAL1 except that, because it is necessary to maintain a strict timing relation between source and destination despite the variable bitrate, it may be necessary to send some of the SAR-PDUs only partially filled.

Once developed, maybe the use of partially filled cells will also become an important means for limiting telephone voice propagation delays as may otherwise have been encountered using AAL1 (see previous discussion).

9.5 AAL Type 3/4 (AAL3/4)

The standards originally foresaw two separate AAL types – 3 and 4, corresponding with the connection classes C and D, as listed in Table 9.2. In the process of development these have become equivalent. They are now simply referred to as AAL3/4.

AAL3/4 provides for packet and frame-based data carriage in either a *connection-oriented* or a *connectionless* mode between AAL-service access points (AAL-SAP) across an ATM network. AAL3/4 can carry variable length packets or frames, and may distribute them in either a point-to-point or a point-to-multipoint manner. In the point-to-point mode two-directional transmission of user data is possible. In the point-to-multipoint mode a unidirectional broadcast transmission is possible between a single source AAL entity and multiple destination AAL entities (Figure 9.4).

Frames of packets of data are split up by the convergence sublayer of AAL3/4 to be carried by one or more ATM cells as necessary. Thus the SAR-PDU which goes to make up the ATM cell payload may be a cell containing either the *beginning of message (BOM)*, the *continuation of message (COM)* or the *end of message (EOM)*, where by 'message' is meant the data frame or packet. This information is coded into the *segment type (ST)* field of the SAR protocol data unit (PDU) header (Figure 9.5).

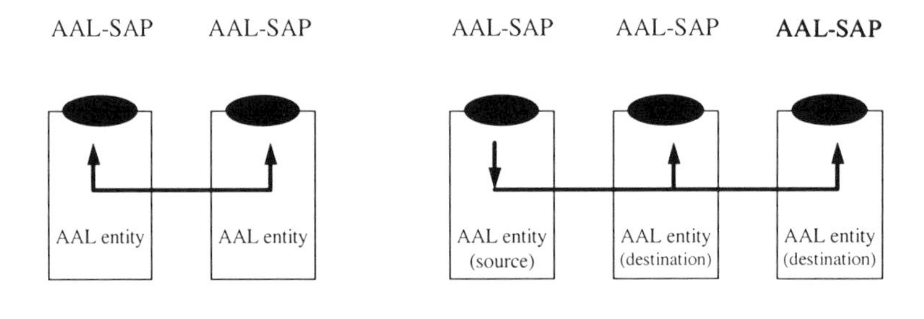

Figure 9.4 Types of AAL3/4 connections

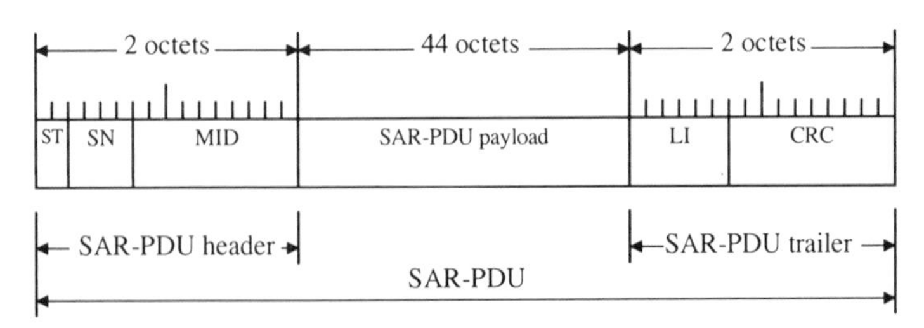

Figure 9.5 SAR-PDU format for AAL3/4

Alternatively, the ST may also be set to *SSM (single segment message)* where a single ATM cell is sufficient to carry the complete data frame, packet or block.

Figure 9.5 illustrates the ATM cell payload format (i.e. the SAR-PDU format) for AAL3/4.

The *MID (multiplexing identification)* field allows a number of different AAL *sessions* (i.e. connections) to share the same ATM layer (VCC) connection. Up to 2^{10} (1024) different sessions may thus share the ATM connection. (The same number of *logical channels* are allowed between an end-user terminal and an X.25 or frame relay network – this is not coincidence, since one of the uses for which AAL3/4 was originally conceived was for X.25 and frame relay.)

For each *multiplexed* AAL connection, a *sequence number (SN)* of 4 bits (modulo 16) is allocated. The sequence number is used to ensure the correct order of receipt of cells.

The *length indication (LI)* field is used to indicate how many of the octets within the SAR-PDU payload (total 44 bytes) are actually being used to carry user information. Where the payload is full (as will always be the case for BOM and COM SAR-PDUs) the LI is set to the maximum value of 44. However, in the case of EOM and SSM payloads the user information will only partly fill the SAR-PDU payload. In this case the LI is set to indicate the actual length of the user information in octets. The minimum permissible value of the LI is 4 (for EOM payloads) and 8 (for SSM payloads).

The *CRC (cyclic redundancy check)* field is a 10-bit error detection code, which is used to detect received errors in the SAR-PDU *header*, *payload* and *length indication* fields. The field is set to the value equalling the remainder of the division (modulo 2) by the generator polynomial of the product of X^{10} and the SAR-PDU (without CRC). The CRC-10 generator polynomial used for the division is $1 + X + X^4 + X^5 + X^9 + X^{10}$.

The *convergence sublayer (CS)* of AAL3/4 provides for the conversion of variable bit-rate data streams into a format suitable for segmentation and reassembly by the SAR of AAL3/4. As already discussed it is subdivided into a *common part convergence sublayer (CPCS)* and a *service specific convergence sublayer (SSCS)*.

The *CPCS* provides for transfer of data frames with any length between 1 and 65 535 octets, in an *unassured* manner. *Unassured* means that there is no guarantee of correct receipt of all data. While the CPCS provides for error detection and indication, there are no error recovery and resending techniques as, for example, provided by X.25 protocol. These would be provided by a SSCS for X.25 (were there one).

The number of CPCS connections which may be established between two endpoints is 1024, the same number of logical channels available in both X.25 and frame relay.

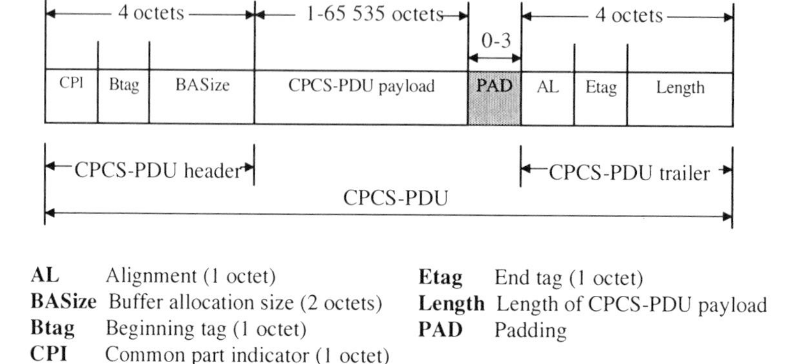

AL	Alignment (1 octet)	**Etag**	End tag (1 octet)
BASize	Buffer allocation size (2 octets)	**Length**	Length of CPCS-PDU payload
Btag	Beginning tag (1 octet)	**PAD**	Padding
CPI	Common part indicator (1 octet)		

Figure 9.6 CPCS-PDU format for AAL3/4

Since frames carried by the CPCS of AAL3/4 may contain up to 65 535 octets, it is clear that they have to be split up and carried by a multiple number of SAR-PDUs (each carrying up to 44 octets of the message). The frames are carried in a CPCS-SDU (service data unit), carried as the payload of a CPCS-PDU, as shown in Figure 9.6.

The *Common part indicator (CPI)* is a field which may be used by different service specific sublayers to indicate different meanings of the CPCS-PDU header information, specifically the units for the values in the BASize and length fields.

The beginning (Btag) and end tags (Etag) are set to the same value in order to ensure that CPCS-PDU headers and trailers (which are carried by separate SAR-PDU cells) are correctly associated with one another. The value is incremented for each successive CPCS-PDU.

The *buffer allocation size indication (BAsize)* field informs the receiving convergence sublayer *entity* of the maximum buffering requirements which will be required to receive the CPCS-SDU.

The CPCS-PDU payload is required by the specifications to be an integral multiple of 32 bits (4 octets) in length. For this reason, a *padding (PAD)* of up to 3 octets of 0's may be added to the end of the user information.

The *common part convergence sublayer (CPCS)* of AAL3/4 supports both class C and class D adaptation layer service. In the class C mode it provides a frame relaying telecommunication service, but requires to be used in conjunction with a *service specific convergence sublayer* (e.g. the *frame relaying service specific convergence sublayer (FR-SSCS)* as specified by ITU-T recommendation I.365.1. Used as a *connectionless network access protocol (CLNAP)* in the class D mode, there is no need for a service specific convergence sublayer.

9.6 AAL Type 5 (AAL5)

AAL5 was developed later than AAL3/4 to serve a similar purpose but in a simpler and more efficient manner. The main functionality is similar, but the multiplexing capabilities of AAL5 are not as powerful as those of AAL3/4.

In AAL5, the *segmentation and reassembly (SAR)* layer protocol information is kept to a bare minimum. The 3-bit *payload type (PT)* field of the *ATM layer cell header* is used to indicate the type of cell. Thus instead of using part of the ATM cell payload to indicate *BOM, COM, EOM*, etc., this information is instead carried by the PT field. The equivalent information is carried within the *ATM-layer-user-to-user (AUU)* parameter (see Table 8.2). This is a parameter value carried by the ATM layer on behalf of the ATM layer user. AUU = 0 is set to indicate BOM (beginning of message) or COM (continuation of message). AUU = 1 is set for EOM (end of message).

The multiplexing, length and error detection information used by AAL3/4 is not mirrored by AAL5. Instead, these functions are left for the higher layer protocols to perform as needed.

As a result of the much greater simplicity of the *segmentation and reassembly (SAR)* of AAL5, much greater line efficiency can be achieved, since now 48 octets of each cell (rather than only 44 as with AAL3/4) are available for carrying the CPCS-PDU as a payload (Figure 9.7).

The *common part convergence sublayer (CPCS)* of AAL5 also performs a similar function to that of AAL3/4, and is also simpler than its AAL3/4 counterpart. The protocol information of the CPCS in AAL5 is all packed into an 8 octet *trailer*, as shown in Figure 9.8. Note that the data frames which may be carried by AAL5, like AAL3/4, can be of any length from 1 to 65 535 octets.

The data block to be carried (e.g. frame) is carried in the CPCS-PDU payload. A padding (PAD) of between 0 and 47 octets follows it to make sure

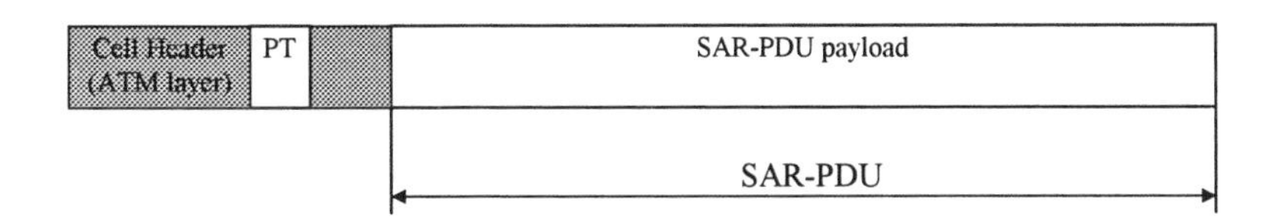

PT Payload Type (3 bits of the ATM layer cell header)

Figure 9.7 SAR-PDU format for AAL5

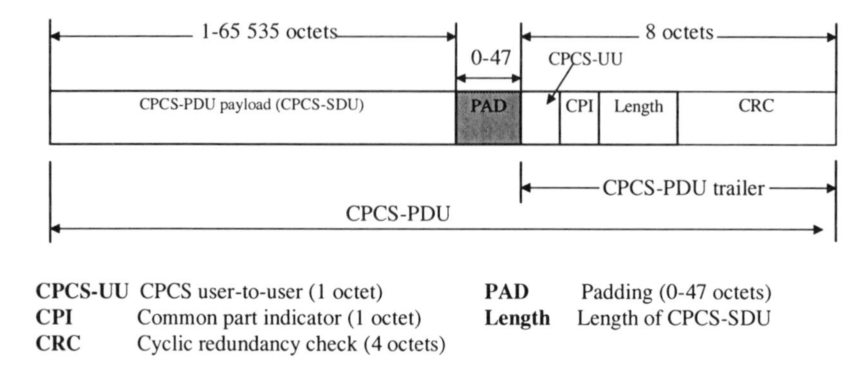

CPCS-UU CPCS user-to-user (1 octet)
CPI Common part indicator (1 octet)
CRC Cyclic redundancy check (4 octets)
PAD Padding (0-47 octets)
Length Length of CPCS-SDU

Figure 9.8 CPCS-PDU format for AAL5

that the CPCS-PDU has a size equal to an integral number of AAL5 SAR-PDUs (i.e. cells).

The *CPCS-UU (CPCS user-to-user indication)* field is used to carry certain information transparently on behalf of the higher layers (e.g. a service specific convergence sublayer).

The *CPI (common part indicator)* field is planned for use in conjunction with layer management functions.

The *length* field indicates the length of the CPCS-PDU payload field (i.e. the size of the user data block). It may be used as a check by the receiver to detect loss or gain of information during transmission.

Finally, the *CRC (cyclic redundancy check)* is a 32-bit CRC error detection code for detecting errors in the remainder of the CPCS-PDU. It is a standard CRC-32 code. For those interested in being able to set or check its value, the generating algorithm is given by ITU-T recommendation I.363. It is the same as that used in Ethernet and Token Ring LAN protocols and in FDDI.

9.7 Service Specific Protocol Support by ATM Networks

Many telecommunications end devices communicate over ATM networks by their direct use of the ATM protocol stack, including relevant AAL. However, for devices not specifically designed to operate over ATM networks, the ATM adaptation layer (AAL) acts as a conversion medium, and in this case the common part AALs (AAL1, AAL2, AAL3/4 and AAL5) need to be supplemented by an extra *service specific convergence sublayer (SSCS)*.

In its simplest form, an SSCS might provide for the simple *encapsulation* of information for carriage by the most appropriate common part AAL. The *Internet Engineering Task Force (IETF)*, for example, has adopted this approach in some of its standardization work, thereby producing a number of relatively simple solutions for LAN interconnection (*routing* and *bridging*) by means of ATM networks.

9.8 Encapsulation

Encapsulation may be used for *bridging* LAN or LAN-type networks (e.g. FDDI networks). In the encapsulation process, the OSI layer 2 protocol of the LAN (i.e. the *logical link control, LLC*) is simply packed into the payload of an AAL5 or AAL3/4 CPCS-PDU payload, and unpacked at the destination. The two end devices then appear as if they were connected to the same LAN. This is all very well if you can afford the bandwidth between the two ends (say, on a campus) but not very efficient for *wide area network (WAN)* use.

Encapsulation is also possible for *router* connection of LANs. A *router* connection is generally more efficient than a *bridge* connection, since only relevant information is *routed* across the wide area network (WAN) using a carefully chosen *route*. But even this solution wastes the routing potential of ATM, since the *routing* (i.e. the choosing of the route through the network) needs to take place in *routers* which are user end equipment as far as the ATM network is concerned. As Figure 9.9 illustrates, encapsulation allows existing LANs to be interconnected across ATM networks using *routers*. The

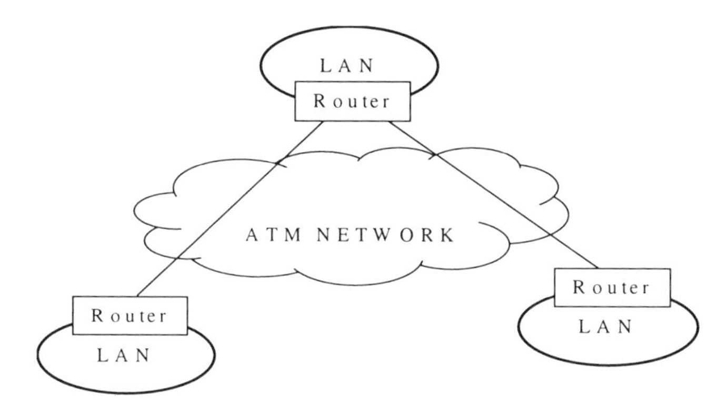

Figure 9.9 Router interconnection across an ATM network using *encapsulation*

drawback is that the ATM network may have to be traversed several times, once for each inter-router connection.

9.9 VC-based Multiplexing

A slightly more efficient router network than that achieved by Figure 9.9 is achieved by *VC-based multiplexing*. This is also a form of *router* interconnection across an ATM network specified by the IETF. In VC-based multiplexing the routers are fully meshed by virtual connections supplied by the ATM network, thereby avoiding the need for intermediate routers. The LAN bridge or routing protocol is encapsulated for transmission between the two ends in much the same way.

Unimpressive, you may think – adding string and sealing wax to the most advanced technology to bring it down to the level of its forefathers. But before we condemn the approach let us consider how useful ATM would be if users' existing devices could not communicate by means of it. It would be like trying to sell the world's first telephone to a customer who would have no-one to talk to!

9.10 The Service Specific Convergence Sublayer for Frame Relay (FR-SSCS)

ITU-T, like IETF, has also defined a number of specifications for the interworking between existing protocols and ATM or for their carriage over ATM. In the case of *frame relay* (or correctly *frame relaying*), ITU-T recommendation I.365.1 defines a service specific convergence sublayer (SSCS) enabling ATM networks to support frame relaying protocol to end devices. This is based on the common part provided by AAL5. The recommendation defines how the protocol functions defined by ITU-T recommendation Q.922 (frame relay) can be carried by AAL5. The Q.933 frame relay status signalling may thus be run equally as well over an ATM network supporting I.365.1 and AAL5 as over a native frame relay network supporting the normal logical link functions defined by Q.922.

Figure 9.10 illustrates how a frame relay terminal, connected to a native frame relay network, may communicate with a frame relay device connected to an ATM network. The Q.933 frame relay status signalling passes directly between the ends. The Q.922 protocol, however, is adapted by the interworking device (possessing the *interworking function, IWF*) to the ATM format, using FR-SSCS. This conversion is defined in detail by ITU-T recommendation I.555.

The *interworking function (IWF)* of Figure 9.10 provides for the conversion

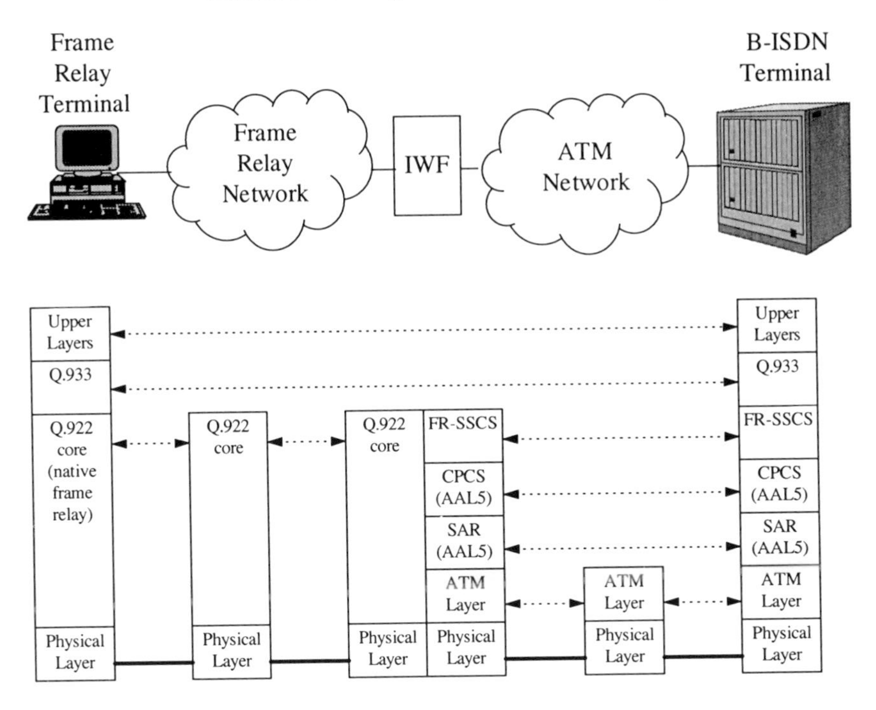

Figure 9.10 Interworking of frame relay and ATM networks according to ITU-T I.555

of protocols between the native core functions of frame relay (equivalent to the *user plane*) as defined in ITU-T Q.922 and the alternative provided by the combination of FR-SSCS, AAL5 CPCS, AAL5 SAR and ATM layer protocols. The upper layer information (including the user information) is oblivious of the lower layer protocol conversions necessary for the information transport.

9.11 Switched Multimegabit Data Service (SMDS) Carriage over ATM

As for frame relay, the ATM protocol suite has also been extended with specific protocols to allow for the easy carriage of *switched multimegabit data service (SMDS)* by ATM networks. These will be of particular importance in the United States where SMDS as a service has already gained some footing. They will allow recently purchased SMDS end equipment to be used in conjuntion with emerging ATM networks. They will, however, be of less relevance in Europe, where SMDS has not been widely accepted or introduced, and the market is instead likely to jump straight to native ATM protocols for broadband applications.

The service specific protocols required for SMDS are defined in the ATM forum Inter-carrier interface (B-ICI) specifications. They define how the *SMDS interface protocol layer 3 (SIP3)* may be adapted for carriage by AAL3/4.

9.12 Service-Specific Connection-Oriented Protocol (SSCOP)

The *service specific connection-oriented protocol (SSCOP)* is a service specific convergence sublayer protocol for the *assured* or *unassured transfer* of data. The protocol is specific to the ATM, but offers data transfer similar to OSI layer 2 (HDLC). As we saw in chapter 6, the SSCOP requires AAL5 as the next lower layer, and itself is used on the *control plane* in association with the *service specific coordination functions (SSCF)* defined in ITU-T recommendations Q.2130 and Q.2140 to support the higher layer signalling protocols (DSS2 and B-ISUP/MTP) which are used at UNI and NNI. SSCOP is defined by ITU-T recommendation Q.2110.

The primary function of the SSCOP is to ensure, using an end-to-end protocol, that all frames arrive successfully and in the right order. This is achieved by using *data sequence numbers*, *poll* messages and *status* message. Each frame of data is transmitted by the sender in conjunction with a *data sequence number (DSN)*. The SSCOP forms the lower sublayer of an SSCS created specifically for ATM network signalling. The upper sublayer is the SSCF, which adapts the format to the specific needs of DSS2 (at the UNI) or B-ISUP (at the NNI). Originally only a single *SAAL (signalling ATM adaptation layer)* was defined, comprising both the functions of SSCOP and SSCF. The sublayering was introduced later to allow maximum common functions but different services to the higher layer. (The DSS-2 protocol is asymmetric with a *user-* and a *network-side*. The B-ISUP protocol assumes two equal partners.)

The use of Q.2931 and SSCOP/SSCF (also called SAAL in combination) as opposed to the forerunner of Q.2931, Q.93B and the original SAAL, is the prime difference between ATM Forum's UNI v3.0 and UNI v3.1. The two are thereby incompatible in one main aspect – the signalling and setup of ATM *switched virtual circuits*. In other respects UNI v3.0 and v3.1 are compatible.

Data frames are not individually acknowledged by the receiver. Instead, the sender is required periodically to *poll* the receiver. The *poll* message requires the receiver to respond with a *status* message. In the status message the receiver informs the sender of the expected sequence number of the next data frame it will receive, and the numbers of any frames it has not received. This allows the sender to repeat any missing frame. Since only the missing frames (and not all subsequent frames as well) are retransmitted, the protocol is efficient in its use of the line capacity, even in this *assured* transmission mode (Figure 9.11.)

The *poll* and *status* messages also serve as a heartbeat mechanism, allowing both sender and receiver to be sure that the other party is still active – that the line and all equipment are functioning properly. Should either the receiver not receive poll messages or the sender not receive status messages in response, then the connection will be cleared.

9.13 Connectionless Services

In a connectionless service, there is no direct acknowledgement returned to the sender during the period of transmission to confirm correct receipt of the information. In effect, the sender transmits the message or other information into a black hole. The *Internet*, for example, provides a connectionless service, allowing senders of electronic mail messages to post their messages into the

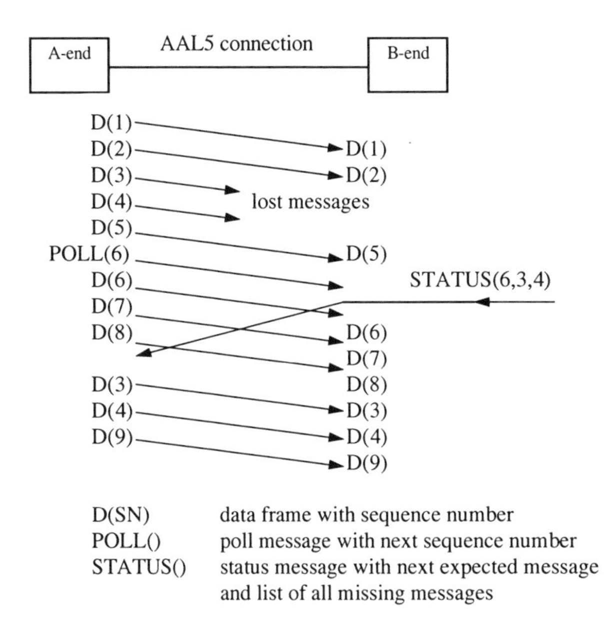

Figure 9.11 Basics of the service specific connection-oriented protocol (SSCOP)

network, without confirmation of their arrival (and indeed with no guarantee that they will arrive).

A connectionless protocol allows data to be transmitted by the sender, and may, through sequence numbers and error check mechanisms, allow the receiver to detect and maybe even correct errors (e.g. wrong arrival sequence of frames). Unlike a connection-oriented protocol, no connection is established, so for example there is no guarantee that the receiver is ready to receive the information. If you don't get a reply to the electronic mail you sent over the *Internet*, you may not know whether this is because your own message did not arrive or because the recipient was too lazy to reply.

The AAL5 protocol, used without any further service specific convergence sublayer, is a connectionless protocol. So is AAL3/4. But do not be misled – the fact that the protocol at the adaptation layer is *connectionless* does not mean that there is no connection at the ATM layer or that there is no physical connection. Instead it means that the sending AAL protocol is unable to tell whether the receiving AAL is listening. Alternatively, there may be some type of store-and-forward mechanism used in the network, by which the information is relayed to the final destination.

10

Operation and Maintenance (OAM) of ATM Networks

The quality of service offered by a telecommunications network depends to a great extent upon the reliability of individual components, but most important is the design of the network as a whole. Since very few networks are subjected to constant and predictable traffic loads, one of the most challenging tasks of managing modern telecommunications networks is the monitoring and maintenance of the network. In this chapter we review ITU's concepts of 'quality of service (QOS)', 'network performance (NP)', performance monitoring and the parameters defined for measuring them. We then go on to describe the functions offered for the monitoring, fault localization and maintenance of ATM networks.

10.1 Quality of Service and Network Performance

ITU-T draws a strict distinction between what it calls *quality of service (QOS)* and *network performance (NP)*. The *quality of service* is a direct measure of the perception of the (usually human) end user of the performance of the system as a whole (network and end equipment). *Network performance (NP)*, on the other hand, is the measure of the performance of the network part of the system alone. Similar parameters may be used to measure both *quality of service* and *network performance* (e.g. propagation delay, *bit error rate (BER)*, per cent congestion, etc.). Normally the measured *quality of service* is lower than the measured *network performance*. The difference is due to the performance degradations caused by the user's own end equipment (Figure 10.1)

The measured quality of service will differ greatly from the measured network performance values where a connection is composed of a number of connections, traversing several different networks and end-user equipments.

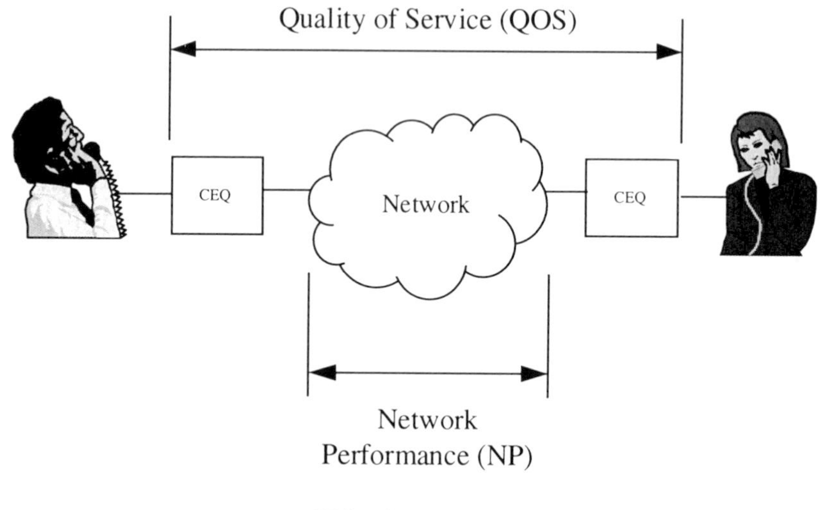

Figure 10.1 The relationship between *quality of service (QOS)* and *network performance (NP)*

While it is of utmost importance to the end user, the problem with *quality of service* as a performance measure is that it is difficult to measure, and would need to be measured for each individual customer separately. This is the reason for the development of the concept of *network performance*. *Network performance* can be more easily measured within the network, and provides for meaningful performance targets for the technicians and network managers operating the network.

In short, QOS parameters are user-oriented, and provide useful input to the network design process, but they are not necessarily easy to translate into meaningful technical specifications for the network. Network performance parameters, on the other hand, provide a directly usable technical basis for network designers and operational managers, but may not be meaningful to end users. Sometimes, the same parameters are appropriate for both QOS and NP, but this is not always the case.

10.2 Quality of Service Parameters for ATM Applications

Quality of service parameters focus on the user's likely perception of the service, rather than upon the technical cause of specific degradations of service. For this reason, different quality of service parameters should be tailored for each different type of network application.

QOS parameters are not comprehensively defined by ITU-T standards. For certain types of services (e.g. leaselines, telephone service, etc.), ITU-T defines QOS parameters, and these same parameters should be used to measure end-to-end quality of such services, independent of what type of transport network is used. Thus, for example, the quality of service of a digital leaseline could be measured in terms of the *bit error rate (BER)*, the *jitter*, the accuracy of the *bitrate*, the propagation delay, the *availability* of the connection (the percentage of a one-month or three-month period for which the circuit was not out-of-service) and the accuracy of the invoice.

Where the leaseline service was provided to the end customer as a *constant bit rate (CBR)* service provided by means of an ATM network, QOS parameters similar to those of a normal leaseline are expected by the customer. But in addition, the network operator has an interest in conceiving network performance parameters which would help him to meet the user's expected QOS level. In this respect, he has a little help, as the ITU-T has defined a number of standard network performance parameters for ATM networks. The network operators task thus becomes relating these to the specific expectations of QOS of individual services, and setting appropriate network performance targets.

10.3 Network Performance Parameters for ATM Networks

The basic types of network performance (NP) parameters to be used in conjunction with ATM networks are defined in ITU-T recommendation I.350. This recommendation defines a set of three different types of NP parameters, and recommends that these should be used as measures of three different functional aspects of the ATM connection (*access* (connection setup), *user information transfer* and *disengagement*). Table 10.1 shows the nine different types of network performance parameters which result. These are termed the *generic primary performance parameters.*

Examples of specific primary performance parameters within each of the generic performance parameter classes are given in Table 10.2. A subset (perhaps all) of these parameters should be regularly measured by ATM network operators (as *network performance* measures). The target values for the chosen parameters need to be set in order to ensure relevant customer satisfaction with respect to *quality of service*. Where no direct relationship exists between quality of service and network performance parameters, experience will help determine acceptable threshold values for NP parameters.

Table 10.1 The *generic primary performance parameters* for measuring NP of ATM networks

	Performance criterion		
Function	Speed	Accuracy	Dependability
Access	Access speed	Access accuracy	Access dependability
User information transfer	Information transfer speed	Information transfer accuracy	Information transfer dependability
Disengagement	Disengagement speed	Disengagement accuracy	Disengagement dependability

Table 10.2 Example primary performance parameters relevant to ATM networks and services

	Performance criterion		
Function	Speed	Accuracy	Dependability
Access	Connection setup delay	Incorrect setup probability (misrouted connection ratio)	Probability of setup denial (connection setup denial ratio)
User information transfer	Successful transfer rate Propagation delay Cell transfer delay Cell delay variation (CDV) Cell transfer capacity	Bit error rate (BER) Cell misinsertion rate (CMR) Cell error ratio (CER) Severely errored cell block ratio (SECBR) Errored seconds	Probability of information loss Cell loss ratio (CLR)
Disengagement	Delay in connection clearing	Premature release ratio Incorrect release ratio	Release failure ratio

In addition to *primary* performance parameters, ITU-T recommendation I.350 also defines the concept of *derived* performance parameters. The most important *derived* performance parameters are those of *availability* and *acceptability*. *Availability* is a measure of the cumulative outage (i.e. non-service) time of the network as a whole (or of a given customer's part of the network) during a given measurement period (e.g. one month or three months).

Availability is measured as the percentage of the total period for which the service was not in outage. Thus the higher the measured *availability*, the lower the network outage. Typical target values are 99, 99.5 and 99.9 per cent.

Acceptability has also been proposed as a derived performance parameter. This would be intended to give a qualitative measure of likely subjective customer opinion of the service level. This gives the potential for inclusion of other more general factors in measuring overall customer satisfaction with the network.

10.4 Operation and Maintenance (OAM) Functions of ATM Networks

In earlier chapters we discussed the various protocol layers of ATM, and in passing, the OAM functions offered by each of the layers. In addition, in our discussion of the ATM *protocol reference model* (PRM) we discussed the concept of *layer management*. This concept laid the basis for the structured management of the various layers and levels of ATM networks.

The ATM standards define five layers of operation and maintenance of an ATM transport network, and label these layers F1 to F5. As we discovered briefly in chapter 8, there are fields reserved in the ATM cell header for coding by the physical and ATM layers, allowing the identification of *OAM-cells (operations and maintenance)*, and in particular for OAM F5 cells (for virtual channel connection management).

Table 10.3 summarizes all the various layer management OAM functions. The table clearly illustrates the carefully layered structure of ATM networks, and the consequent clear segregation of OAM network performance diagnosis and fault finding functions.

If you study the table carefully, you may feel pressed to ask how the sender determines which of the number of different possible fault conditions has led to his receipt of *RDI (remote defect indication* – formerly called *FERF, far end receive failure)*, or how the distant receiver knows which condition led to his receipt of AIS. The answer depends upon whether the AIS or RDI is received as a *physical layer* or *ATM layer* signal. At the physical layer, the relevant message will be accompanied by a *layer management message* indicating the cause of the failure (e.g. LOC, loss of cell delineation, or LOM, loss of OAM cell). At the ATM layer, both AIS and RDI messages (both are types of *fault management cells* at the ATM layer) contain a *failure type* and a *failure location* field.

In addition to providing for fault management, OAM cells at the ATM layer also provide for performance management, and for activation/deactivation of various network diagnostic tools (e.g. network continuity check or cell *loopback* facility – we discuss these capabilities in more detail later in the chapter).

The general format of ATM Layer OAM cells is shown in Table 10.4.

Table 10.3 OAM functions of ATM networks

Layer	Level	OAM function*	Function description
Physical Layer	Regenerator section level (F1)	Signal detection	The signal detection function ensures that the regenerator section is still live (i.e. not out-of-service). If no signal is received at a given point then an AIS (alarm indication signal) is transmitted to further receiving points in the connection and an RDI (remote defect indication, formerly known as FERF, far end receive failure) will be returned to notify the sending end.
		Frame alignment	Loss of frame alignment (LOF) as detected at an intermediate regenerator will also result in an AIS signal being sent to further points along the transmit path and return of RDI to the sending end.
	Digital section level (F2)	Section error monitoring	Should a given bit error ratio (BER) be detected on a given digital section, then AIS and RDI signals are generated.
	Transmission path level (F3)	Cell rate decoupling	Cell rate decoupling is the failure to insert or suppress idle physical layer cells from being passed to the ATM layer. (The idle cells exercise the digital line when no ATM layer cells are waiting to be sent.) The appropriate reaction to this state is not yet fully defined.
		PL-OAM cell recognition	In the case that PL-OAM (physical layer OAM) cells are no longer detected, RDI signal should be returned to sender. (Periodic PL-OAM cell generation is an obligation of the sender.)
		Cell delineation	If cell delineation (i.e. synchronization) is lost then the RDI signal is used to notify the sender.
		CN status monitoring	When the sending customer network does not monitor as OK then the receiver is informed by means of AIS.
		AU pointer operation	The AU (administrative pointer) is a requirement of SDH-based line systems. When absent, the sender is informed by means of the RDI signal.
ATM Layer	Virtual path level (F4)	Monitoring of path availability	When a break in a virtual path is detected by the ATM layer (by interpretation of the failure indications passed to the AM layer by the physical layer) then other points along the VP are notified by means of VP-AIS and VP-RDI signals. These enable the ATM layer, if

Table 10.3 (*continued*)

Layer	Level	OAM function*	Function description
			appropriate, to re-establish the VP using an alternative physical route.
		Performance monitoring	Performance monitoring of virtual paths is achieved by sending regular performance monitoring cells after each block of user cells. These are used to ensure that performance of the VP remains within acceptable threshold values.
	Virtual channel level (F5)	Monitoring of channel availability	VC-AIS and VC-RDI are sent to receiving and sending endpoints of a VC to notify of failure conditions arising from the failure of the connection. They are sent at a nominal rate of one per second until the failure condition subsides. These signals in principle notify the ATM VCC endpoints of the connection failure. Response could either be to attempt an alternative connection, or wait until recovery.
		Performance monitoring	The performance monitoring function of the VC level of the ATM layer operates by sending performance monitoring cells on an end-to-end basis after each block of N user cells. This function can be used to monitor end-to-end errored blocks, misinsertion, cell transfer delay and other performance parameters on a specific connection.

*Performance monitoring or fault localization

The format and information content of the *function specific field* of the ATM layer OAM cell depends upon the OAM cell and function type. In the case of AIS or RDI *fault management cells*, this field contains the *failure type* and *failure location* fields plus a number of unused octets.

OAM activation/deactivation cells are used to activate or deactivate performance monitoring, network continuity checks or diagnosis procedures such as cell loopback.

When performance monitoring is activated, OAM *performance monitoring cells* are generated and sent over the connection to be monitored, in order to gain a real-time assessment of the network performance level. Performance

Table 10.4 ATM Layer OAM cell format

	← OAM cell information field →				
Header	OAM Type	Function Type	Function specific field	Reserved field	EDC (CRC-10)
5 octets	4 bits	4 bits	45 octets	6 bits	10 bits

Header – coded in the normal manner for UNI or NNI cells. Two pre-assigned VCIs are used to distinguish OAM cells. Two pre-assigned PTIs (payload type identifiers) are also used for certain OAM cells.

Function specific field – format depends upon the OAM and function type.

EDC – error detection code – a CRC-10 for detecting errors in the remainder of the OAM cell information field.

OAM type cell (name)	4-Bit OAM type code	Function type (name)	4-Bit function type code
Fault management	0001	AIS (alarm indication signal)	0000
	0001	RDI (remote defect indication)	0001
	0001	Continuity check	0100
Performance management	0010	Forward monitoring	0000
	0010	Backward monitoring	0001
	0010	Monitoring/reporting	0010
Activation/deactivation	1000	Performance monitoring	0000
Activation/deactivation	1000	Performance monitoring	0000
	1000	Continuity check	0001

monitoring cells contain in their specific information fields a number of different types of counters, time stamps, block error detectors, cell sequence number counters and other types of information which allow for extensive performance assessment of the ATM layer in the following terms:

- *cell error ratio (CER)*
- *cell loss ratio (CLR)*
- *cell misinsertion rate (CMR)*
- severely errored cell block ratio (SECBR)
- *cell transfer delay (CTD)*
- *mean cell transfer delay*
- cell delay variation (CDV)

Table 10.5 shows the specific fields of an ATM layer OAM performance management cell.

A continuity check cell is sent *downstream* by a VPC or VCC endpoint when no user cell has been received for a predetermined period, despite no indication of connection failure. If, within a further predetermined period, still no cells are received, then RDI is returned – to inidicate loss of connection continuity (i.e. break in the connection).

The ATM layer loopback facility allows cells to be sent over a given section of a VPC or VCC in the transmit direction and then to be *looped back* and returned over the receive direction of the same connection *segment*. Such a facility is also useful for maintenance personnel using external test equipment for measuring network continuity and performance, including delays and error rates.

The *loopback* principle of checking connections has its roots in the maintenance of circuit switched telephone networks, whereby a single technician at one end of a circuit can test both transmit and receive directions by first arranging for the circuit to be *looped back* at the far end. Typically the

Table 10.5 Specific fields in the ATM Layer *performance management cell*

← 45 octets →						
MCSN	TUC	BIP-16	TS (Optional)	Unused	Block error result	Lost/misinserted cell count
8 bits	16 bits	16 bits	32 bits		8 bits	16 bits

Field	Field name	Purpose
MCSN	Monitoring cell sequence number	Sequence number, modul 256 of OAM monitoring and reporting cells. This can be used as a measure of cell loss and of cell misinsertion (i.e. out-of-sequence).
TUC	Total user cell number	Indicates the total number of user cells, modulo 65 536, since the last monitoring cell. This is used to measure cell loss.
BIP-16	Block error detection code	A block error detection code, calculated over all the user cells transmitted since the last monitoring cell. This is used to measure cell errors and severely errored blocks.
TS or TSTP	Time stamp	When used, the time stamp can be used to measure cell delay and cell delay variation.
BLER	Block error result	This field is used for backward reporting of parity errors in the incoming BIP-16 code.
	Lost/misinserted cell count	This field is used for backward reporting of the number of lost or misinserted cells counted since the last monitoring cell.

technician would use a tone generation device to send a signal on the transmit path. He could then measure on the receive path the frequency, loudness and distortion of the returned signal on the receive path. By doing so he could localize circuit breaks (failures or discontinuities) and make adjustments to improve performance. The disadvantage of the loopback method in circuit switched networks was, however, that the circuit needed to be temporarily taken out of service. With *ATM layer loopback*, OAM cells can be looped back without affecting user cells – the connection can remain in service despite the ongoing maintenance and diagnosis work.

10.5 ATM Networks and *TMN (Telecommunications Management Network)*

The operations and maintenance functions defined for ATM networks are designed according to the principles of the *Telecommunications Management Network (TMN)*. This is a concept and series of technical standards defined by ITU-T which allow a wide range of different types of telecommunications networks and devices to be managed by a network of multiple management systems.

By networking the management devices and standardizing a means for sharing information between them, the various components of the network as a whole can be better coordinated. Failure alarms, for example, can be more easily consolidated and analysed for their root cause. Where isolated network management systems are used for the individual network components, a transmission line failure will result in alarms being generated by the network management systems of both transmission components and switching components. The human operator is then left to consider the various alarms and work out for himself whether there is a single cause of the multiple alarms, and if so, what it is.

The key element of TMN is the Q3-interface used between the network component and its network management system, and the *CMIP (common management interface protocol)*. These define a number of standard procedures allowing the network management system to monitor and control the network element. Specifically, the ATM standards define the following management functions:

- performance monitoring;
- defect and failure detection by continuous monitoring and generation of alarms as appropriate;
- network and system reconfiguration, allowing, for example, changeover to backup facilities at times of network failures;
- fault localization.

In addition to TMN interfaces, some (early) ATM equipment is likely to offer the *SNMP (simple network management protocol)*, since this is the basis of the *interim local management interface (ILMI)*, defined by the ATM Forum for use at the UNI. ILMI enables the management of UNI *managed objects* (switches, devices, parameter settings, etc.) which correspond to functions at the UNI and may be controlled across it. Together the *managed objects* make up the UNI *management information base (MIB)*.

The short term attraction of SNMP as the basis for ILMI lies in its widespread deployment in existing corporate data networks, routers, LANs (local area networks) and *Internet* components. SNMP was developed by the *Internet Engineering Task Force (IETF)* and is defined in its specification RFC 1157.

Figure 10.2 illustrates not only ILMI but also the other network management interfaces being developed by ATM Forum. As well as ILMI (which is defined within ATM Forum's UNI v3.1), the M3– and M4–interfaces (for customer network management and network operator management of public ATM networks respectively) are also published (ATM Forum af-nm-0019.0000 and af-nm-0020.0000 respectively – published October 1994).

Future standardization work is likely to concentrate on migration to the *CMIP* protocol (probably by integration of certain SNMP procedures into it). In addition, a huge amount of work must be applied to the definition of the standard *MIB (management information base)* of *managed objects* relevant to ATM networks and devices. Without these definitions, it will be impossible to develop the necessary tools for management of the network as a whole. At the moment these are poorly defined.

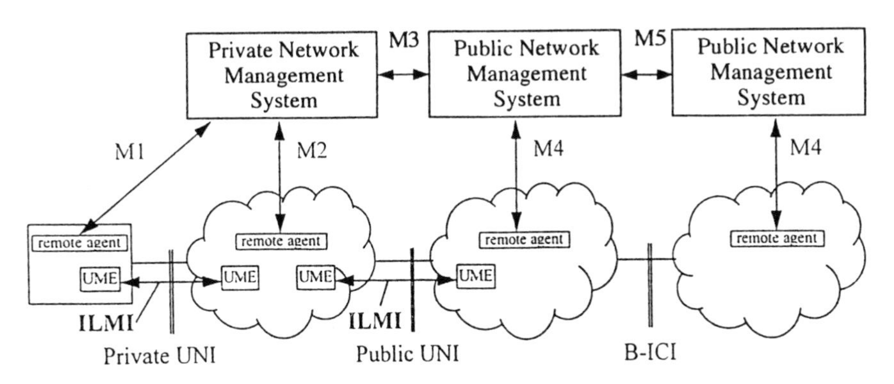

Figure 10.2 Management interfaces in ATM networks (ATM Forum)

11

The Likely Teething Problems of ATM

We have heard how ATM is one of the most promising general purpose telecommunications transport technologies. We have discussed in detail the strengths of the cell switching technique at its core, and the incredible flexibility afforded to ATM transport networks by the ATM adaptation layer. The challenge now for equipment manufacturers and network planners alike is the realization of this promise. In this chapter we discuss some of the problems they will come up against.

11.1 Stabilization of the ATM standards

As with all other telecommunications standards, the establishment and stabilization of adequate standards is crucial to the success of a technology, where this technology requires that networks be built up from various different switching elements and end devices, supplied by different manufacturers.

While the first standards may be technically adequate to allow the first equipment to be developed by individual manufacturers and for it to work, they are often not sufficient to ensure compatibility of equipment produced by different manufacturers. The problem is that it is usually not possible at the first attempt to write the standards in such a way that only a single interpretation is possible. In addition, it is often not until the equipment development phase that the need for standardization of certain functionality or procedures becomes clear. In such cases, manufacturers try to 'interpret' or predict the likely decision of the standards body and implement an interim solution in order that their equipment operates.

The trials of the early users in trying to build networks from equipment supplied by different manufacturers serve as a test for the adequacy of the standards. Usually, such trials lead to further contributions to the standards committees and the resultant clarification of unclear points. It takes several years before all the problems are ironed out – as, for example, in the case of X.25 packet switching. Nowadays one can assume with confidence that X.25 devices provided by different manufacturers will intercommunicate correctly with one another. Such confidence may not yet be assumed with ATM devices.

ATM standards have reached the stage where considerable debate has already led to technically advanced standards, and several manufacturers are already able to offer functional and reliable ATM networks where these are built entirely from their own family of products. Their compatibility with other manufacturers' devices is not initially guaranteed.

The standards for the physical interfaces used in ATM are likely to work reliably, since they are strongly based on digital transmission means and SDH standards which were already available. The basic ATM layer protocol is also likely to operate more or less correctly, particularly at the UNI (user–network interface), since this has been an area for intense study within the standards forums. Other functionality may work less well. Generic flow control, while based on principles developed for frame relay, may not be equally well understood and implemented by all equipment manufacturers. Likewise, the early support of switched VCCs (as opposed to semi-permanent VCCs) may not operate reliably between different manufacturers' devices. The signalling standard needed for switched connections (DSS2), while based upon the narrowband ISDN user-to-network signalling (DSS1), also has a number of new features. The network–node interface (NNI) is also likely to prove more problematic than the UNI, due to its relative youth.

The adequacy and stability of the standards will be initially most critical at the UNI (user–network interface) and at the *INI* (*Inter-network interface*, or *ICI, inter-carrier interface*). It is at the UNI that an end-user device provided by one manufacturer to serve a given ATM application needs to communicate correctly with the switch elements within the network (Figure 11.1).

The INI or ICI will be as important as the UNI as soon as end-users expect to traverse more than one ATM network or sub-network during the course of a connection. This situation arises when the public ATM network in one country is connected with a second public ATM network in the same or another country.

Also the *PNNI (private network–node interface)* will be important as campus and private closed user group ATM networks (e.g. university networks) emerge, with different manufacturers contributing the different components.

The network-node interface (NNI) will first take real significance when major public ATM network operators demand supply of their network

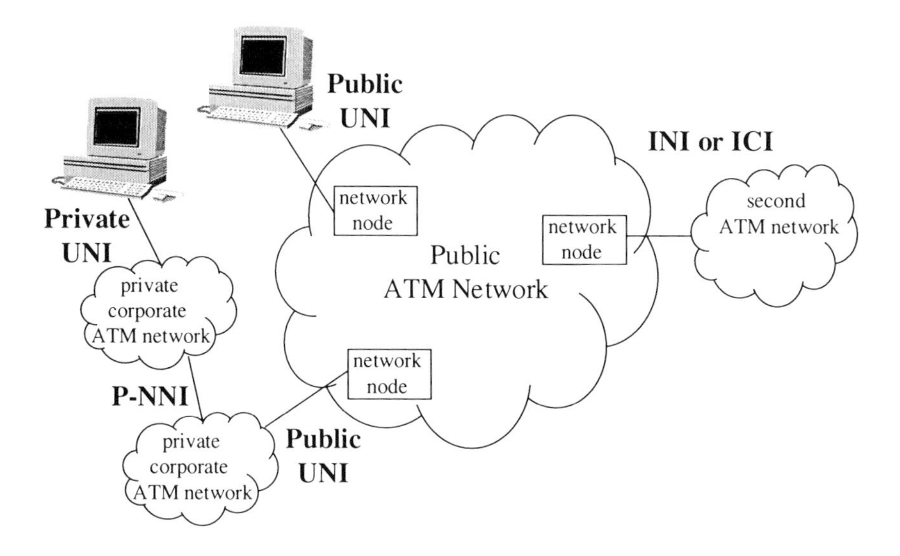

Figure 11.1 The standards most critical to realization of early ATM networks

switching nodes by multiple manufacturers. Then it becomes necessary to specify every aspect of the interconnection of network nodes.

In some early network realizations, ATM switching and AAL functionality will all be provided by the same manufacturer, and the external connections to the network will be based entirely on existing standards such as point-to-point X.21 (constant bit rate) or frame relay (Figure 11.2). Such a network realization has the attraction of allowing the network operator to orientate his current network infrastructure investments around ATM equipment, without exposing himself too greatly to the potential interworking problems associated with different manufacturers' different interpretations of immature standards.

11.2 ATM Hardware for Cell Switching

No ATM network will be possible without hardware capable of the cell switching of ATM connections. As we have seen, ATM is a specialized form of statistical multiplexing or packet switching, in which cells are switched from an input channel to an output channel according to the address information (i.e. the VPI/VCI value) carried in each cell header. In this way, ATM has much in common with earlier techniques such as *X.25 packet switching* and *frame relay*. But while these techniques provide a basis for the technology,

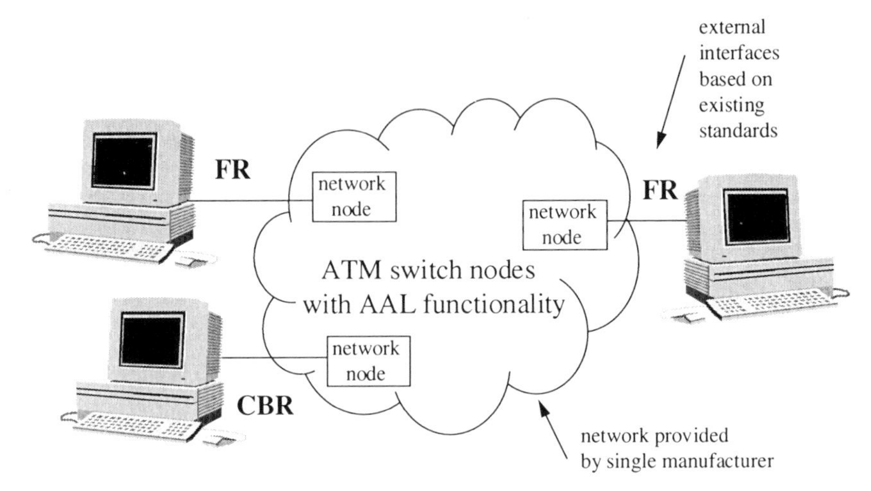

Figure 11.2 Potential early ATM realisation avoiding exposure to immature standards

there is one big difference – the arrival rate of the cells in ATM is much faster. The demands on the ATM switch in processing and reacting to the information in the cell header are much greater than for packet or frame swtiching – much, much greater.

One of the major hardware design considerations facing ATM equipment designers is how to structure the switch to achieve the necessary speeds, and how to arrange the buffer storage which is needed to complement the switching. In the simplest type of switch, a single bus switch, each cell arriving at the switch is broadcast in turn across the bus to each of the output ports. Each output port then extracts from the bus those cells with VPI/VCI addresses relevant to itself (Figure 11.3).

The advantages of the single bus structure of ATM switch (Figure 11.3) are the relative simplicity of the switch structure and the ease with which it is able to support broadcasting (point-to-multipoint connections). The main disadvantage is the relatively high cell switching and bitspeed needed to be supported by the bus (the bitspeed of the bus must exceed the sum of the bitspeeds of all the incoming ports). For larger switches (above about 10 Gbit/s) single bus switches are impracticable.

From Figure 11.3 we also see the need for cell buffers in both switch input and output sides. On the input side, a buffer is needed to temporarily store cells which arrive at the switch input while another cell is already on the bus (cell 4 of Figure 11.3 overlaps cell 3). The buffer stores the cell until the switch bus is free. Buffers may in addition be needed on the output ports, in order to ensure that output ports do not receive more cells simultaneously than they are able to transport away. This situation always arises due to the simple fact that the

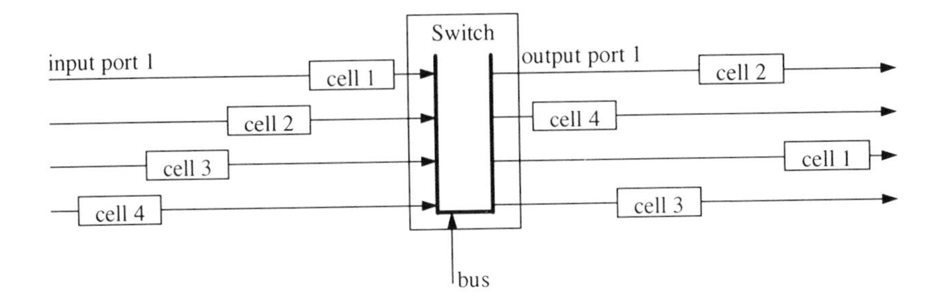

Figure 11.3 Single bus structure of simple ATM switch

bus bitrate is higher than the outgoing port line rate. But over and above this effect, the output buffer must also adjust for instants in time when various input ports send cells simultaneously to the same output port (i.e. cells relating to different virtual connections).

Designing the position of the buffers and their size has turned out to be one of the more difficult aspects of ATM switch design. The buffers need to be large enough to ensure that excess numbers of cells are not lost through buffer overflow. But buffers which are too big can lead to unacceptable cell delays and unacceptable cell delay variation.

Larger ATM switches are generally based on matrix-type crosspoint space switches (Figure 11.4). These allow full availability switches (every input can reach every output without internal blocking, even at very heavy switch load) to be built without requiring impracticable clocking speeds (switch matrix bitrate). Such switches also lend themselves well to production as silicon chipsets using VLSI (very large scale integration), as they have very regular structural pattern. Furthermore, the resulting chipsets lend themselves well to combination with one another to make very large switch matrices (Figure 11.5). The

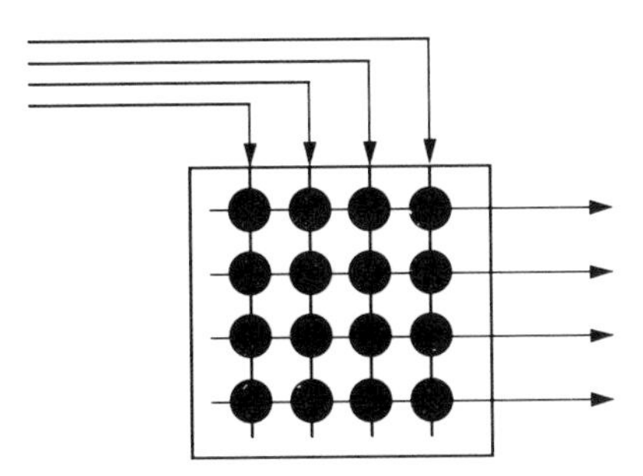

Figure 11.4 4×4 Matrix-type structure of ATM switch

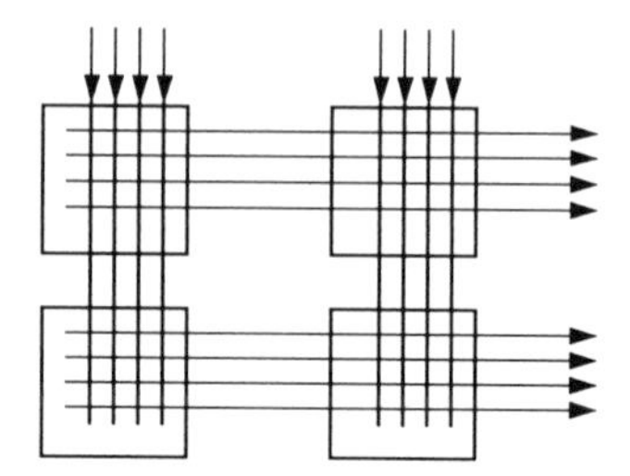

Figure 11.5 Creating an 8 × 8 switch matrix from four 4 × 4 switch chipsets

disadvantage of this type of switch structure is its complexity. Such switches require accurate synchronization and control of the switching mechanism.

11.3 Switching of broadband Data Applications using ATM

The boom in demand for the interconnection of personal computers and *local area networks (LANs)* is creating much of the early demand for high bandwidth data networks, and most of the main computer and LAN manufacturers (e.g. Bay Networks, Cisco, DEC, IBM, Nixdorf, SUN, etc.) are rapidly bringing out ATM communications cards for their devices. In particular, the emergence of client/server computer architectures, the increasing practice of information database sharing (including via *Internet*) and broadcasting of graphic images is creating the need for high-speed data networks.

Data traffic between LAN networks is notoriously *bursty* in nature, usually comprising very short bursts of very high bitrates, punctuated by long idle periods. The bursts follow software commands or the typed requests of human users requesting downloading or posting of particular software or database files. The high speed of the network enables information to be delivered within a short *response time*, but it is critical that it arrives without errors, if necessary at the cost of a longer response time.

Experience with some of the early ATM networks has shown that the burstiness of data traffic has tended to be underestimated, and buffers in the ATM switches were too small. The result was excessive buffer overflow and consequent heavy resending of lost data. User response times (i.e. the effective end-to-end data transfer rate) fell far short of expectations. Certain ATM switch manufacturers have responded to these problems by extending the internal switch buffers. It remains to be seen whether other measures will be necessary or new side-effects need to be corrected.

For those users contemplating ATM networks for a specific high-speed data application, there is no better advice than to run a test to determine what response times will be achieved!

11.4 The Teething Difficulties of *Multimedia*

The clamour for *multimedia* applications is often quoted as being one of the major drivers in creating demand for ATM and broadband-ISDN (B-ISDN) networks, but does anyone really know what *multimedia* is?

The idea behind *multimedia* is that totally new types of devices and computer applications become possible if several different types of information (e.g. text, video, speech, etc.) can be carried over the network between two or more different user devices simultaneously. Thus a company employee, sitting at his PC, might be able, without closing the word processing, interactive video encyclopaedia and electronic mail 'windows' he is currently using, to receive a video telephone call from one of his colleagues in a new window. During their conversation, his colleague might perhaps send him a detailed building plan as a data file and a set of meeting minutes as a fax – for them to discuss together. These could appear in fifth and sixth 'windows'. A seventh window might also be used to bring another colleague into the discussion part way through the call.

Wow! Multimedia opens the door to a brave new future, you might think. And maybe it does, but just before you get carried away you ought to consider for our example above what advantage a multimedia network (such as ATM) has over separate videotelephone, data and fax networks. Of course multimedia ATM networks will allow our company employee to telephone (in video form) and send data and fax messages simultaneously. But in principle there is no reason today why individuals cannot telephone and fax at the same time!

Put straightforwardly, the critical question is 'what can I do with a multimedia ATM network that *I could not previously do* with "multiple network multimedia"?' Current *multimedia* products and applications do not appear to provide an answer which the market buyers find compelling. The compelling reason is likely to take three to five years to emerge. Like ISDN and like the laser before it, *multimedia networking* is largely a technology before its time – a solution looking for a problem. We need not fear that the computer and software manufacturers will be idle in conceiving new possibilities and products (and particularly ones brandishing the ritzy *multimedia* claim), but it will be some time before a mass market for multimedia communications networks is achieved. This will limit initial demand for ATM, and so delay the price reductions that can later be expected.

11.5 Telephone and Voice Switching by ATM Networks

The viability of ATM as an *integrated network* technology, allowing both voice and data signals to be carried efficiently and simultaneously over the same network, will rely largely upon its credibility as a voice switching technology. The capabilities of ATM as a broadband and variable bitrate data switching technology go without question. The cell switching technique used as the basis for ATM is an advanced form of data packet switching.

There are a number of aspects of voice switching which ATM network designers should be aware of and will have to wrestle with:

- the extreme sensitivity of telephone signals to delays, echoes and variable delays;
- the difficulties caused by voice compression;
- the quality problems associated with *silence suppression*;
- the very heavy call control processing demands made by telephone network traffic; and
- the problems of interworking an ATM voice network with an existing telephone network.

We discuss the problems in turn.

Delay

For a human listener in a telephone conversation, a long propagation delay (i.e. transit time between speaker's mouth and listener's ear) is very distracting. A half-second delay, for example, is typical of an international telephone circuit carried via a satellite channel. The distraction is that the speaker, having completed his sentence, will experience at least one second silence before his listener appears to respond, even if the listener responds immediately after himself hearing the sent signal.

Normally in telephone conversation the speaker (party A) expects audible response from the listener (party B) almost immediately, if only a 'hmmm' to indicate that the listener has heard. Many speakers find the pressure of waiting (even as little as one second) too much, and start to question 'are you still there?'. Having started the new sentence, the first reply arrives, so party A stops talking again. Meanwhile party B breaks off from his response to avoid them both speaking at once. A period of silence follows . . . then both start to talk at once again. Intercontinental telephone users (e.g. USA–Australia or Europe–Asia) will relate to this experience.

The circuit-switching technique used by pre-ATM telephone networks incurred only the unavoidable propagation delays caused by the limited speed of electromagnetic transmission (light or electrical signals). By contrast, the cell switching technique used in ATM networks may add markedly to the propagation times of voice or telephone signals.

The first source of added delay for voice signals carried by ATM networks is the size of the cell information payload (48 octets). A normal telephone voice signal of 64 kbit/s is composed of an 8-bit code (1 octet) sent every 1/8000th of a second. In order simply to fill one cell, the sending device will have to wait 48/8000 seconds (6 ms). Insignificant, you might say. And alone it is, but this delay is additional to the propagation delay. Further, there are switching delays added at each ATM switching point in the connection. These can each add delays of similar duration. All insignificant until there are too many links in the chain!

Echo

Once the propagation delay exceeds about 32 ms, *echo* can become a problem. This is usually more offputting than the simple delay. The speaker becomes aware of his own voice echoing back to him. The echo arises because some of the signal created by his voice is reflected either in the electrics of the distant end-user's exchange or in his handset (for those readers to whom it may mean something, this occurs in the *hybrid*, the 2-wire to 4-wire conversion device necessary with analogue telephones). Alternatively, an echo can result when an audible signal is carried from the earpiece to the microphone of his correspondent, though this is usually not the cause.

There is nearly always a signal reflection from the far end (Figure 11.6), but provided the delay is not too great, the speaker is not aware of it. After all, a speaker gets used to the sound of hearing his own voice as he talks!

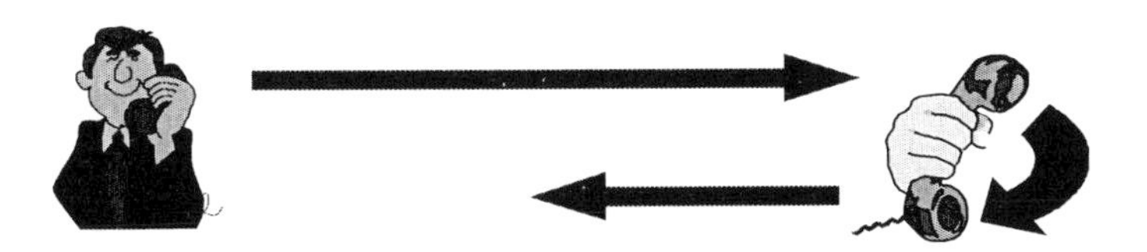

Figure 11.6 The cause of echo in telephone networks

In order to suppress echo, it is usual in telephone networks to build *echo cancellers* into connections where the propagation delay is (or is likely to be) greater than about 32 ms. As the name suggests, echo cancellers remove echoes by electrically negating them. ATM networks carrying voice signals must nearly always be equipped with echo cancellers since the end-to-end propagation delay nearly always exceeds this limit.

Variable delay

As if delay and echo were not challenges enough for ATM network designers and operators, there is also *delay variation* to cope with. When the variation in the voice signal delay becomes too great, the listener perceives the signal as rather staccato and 'chopped up': . . . 'Ick caa see somewhaa uninteh..gible' (It can seem somewhat unintelligible). The avoidance of this problem will rely on very good understanding of the statistics of network traffic and the resultant definition and maintenance of acceptable network loadings.

Voice compression

Voice compression complicates the problems of delay, echo and delay variation. The lower bitrate of compressed voice tends to magnify the problems. In the case of a voice signal compressed to 32 kbit/s, the time to fill an ATM cell is 12 ms instead of 6 ms. At 16 kbit/s (already widely used in private corporate telephone networks) the time to fill an ATM cell is 24 ms. This makes echo cancellation a must on every connection, and makes stringent demands on containing delay variation.

Silence suppression

Silence suppression (Figure 11.7), as we discussed in chapter 4, is a very effective way of reducing the network bandwidth required to carry a telephone conversation. The efficiency is created by not transmitting any signal when the speaker is silent. Since only one person talks at a time, this saves at least 50 per cent of the bandwidth.

Used on telephone calls between two individuals in quiet offices *silence suppression* can be effective and relatively unobtrusive. However, in an open-plan office where there is a humming level of background noise, the effect can be distracting. To the listener, the background noise behind the

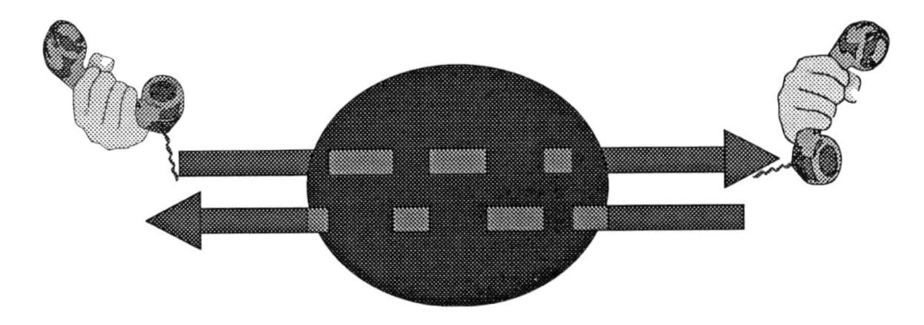

Figure 11.7 Telephone network bandwidth saving by means of silence suppression

speaker (which the system may assume to be 'silence' and therefore will suppress) starts and stops as the speaker talks, or as the listener responds. An added distraction can be that the first part of some of the speaker's words appear to be cutoff. These problems are likely to be relatively quickly overcome by hardware and software improvements, but experience will need to be gained from the first live systems.

Telephone call processing

Voice and telephone traffic will not only make strenuous demands on ATM networks in terms of its quality needs. In addition it also demands exceedingly heavy call setup processing demands. ATM switches will need to be capable of analysing the dialled digits and setting up telephone calls exceedingly quickly, in order that they can cope with the total number of calls to be set up.

Let us consider the simple network of Figure 11.8, in which an ATM switch is connected to only two 155 Mbit/s trunk lines. Let us assume that the lines carry exclusively telephone traffic. Calls originate 50 per cent from the left hand side and 50 per cent from the right-hand side, and simply transit the switch. Let us also assume that the lines are filled to capacity with telephone traffic.

Using the example of Figure 11.8, let us now calculate how many telephone calls the switch must be capable of setting up within an hour (the *busy hour* – when the trunks are used to capacity). Let us assume that the calls are compressed to 32 kbit/s and that *silence suppression* generates a further 50 per cent bandwidth saving. Each telephone call therefore requires a bandwidth of 16 kbit/s. When used to full capacity, the 155 Mbit/s available can therefore carry 9720 simultaneous calls (155 520 000/16 000).

If the *average holding time* (i.e. duration of each of the calls) is 3 minutes, then the number of calls to be setup by the switch per hour is 194400

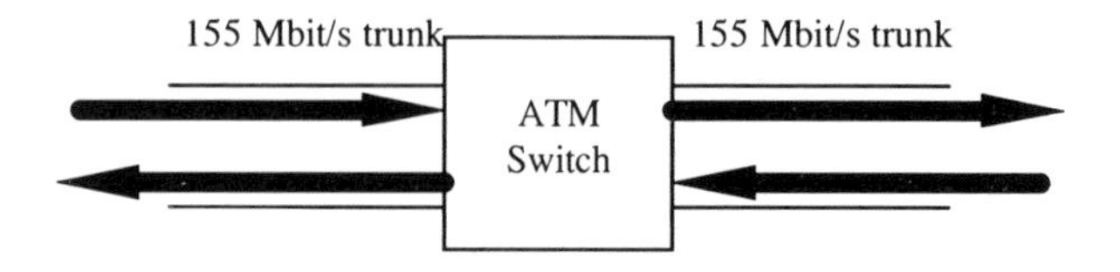

Figure 11.8 An ATM switch carrying transit telephone traffic

(9720 × 60/3). But in addition to the calls which get setup successfully, there are also a number of *call attempts* which do not result in a successful call. Unsuccessful call attempts typically account for around one-third of total *call attempts*. The total *busy hour call attempts (BHCA)* which our switch in Figure 11.8 must therefore deal with is around 300000 (300 kBHCA).

A typical modern public telephone exchange in its maximum configuration is capable of processing 400 kBHCA. The most powerful exchanges cope with around 1200 kBHCA. In other words, if we quadrupled the number of trunks in our exchange (to eight 155 Mbit/s trunks) and continued to use the switch exclusively for telephone traffic then we would need a call setup processor more powerful than is available today. Put another way, the switch will be incapable of setting up the number of calls it needs to in order to fill the trunk line capacity.

In practice, the traffic through early ATM switches will not be entirely voice traffic. The initial demand will be for broadband connections for which less call setup processing is required per unit of bandwidth. In this respect, corporate networks are an ideal application of ATM switching, since they typically require to carry a relatively large proportion of permanent high speed data connections (e.g. for LAN interconnections, videoconferencing and mainframe computer networks) in addition to the internal corporate telephone connections.

ATM is likely to form the basis of the next generation of public telephone networks. But before a wholesale replacement of the existing telephone network can commence, the problem of the lack of call setup processing power must be solved (in order that the existing volume of calls can be coped with). The most tenable next step is probably a technology created by 'grafting' an ATM switch matrix onto a telephone exchange call processor – in effect replacing the telephone circuit-switched matrix used today with an ATM one. Such an approach allows much of the existing telephone network signalling, sophisticated software functionality and high call processing throughput to be reused in an ATM world. In addition, it generates the very tempting prospect of an early combination of the powers of ATM and *intelligent network (IN)* technologies (as we discussed in chapter 4).

11.6 Interworking with Existing Networks and Applications

A common problem for all new network technologies is that potential users are restricted in the number of other users with whom they may communicate. The potential user is thus often faced with relatively high costs of investing in the new system, but then has no-one he can 'talk' to using it. The usual solution is to build a 'backward compatibility' into the new system, allowing him to 'talk' to the users of earlier networks and technologies. In this way, the new user gains the benefits of the new technology when talking to other new network users but does not lose the ability to communicate with those connected only to older networks.

The standards bodies are specifying the means for interconnecting and interworking ATM networks with older network technologies, including frame relay, SMDS, narrowband ISDN and *Internet*. These standards will be important in creating user acceptance for ATM by increasing the connectivity of its early users, but it will take some time before all possible interworking cases are defined adequately.

The interworking standards define how an ATM connected device or user must interpret information received from an older network and how they must code information to be sent to the older network. In addition, the standards define the conversion process which must take place at the interconnection of the two networks.

Figure 11.9 illustrates the interworking of an ATM (B-ISDN) terminal with a frame relay terminal in an existing frame relay network. In particular it shows the protocols which must be supported by the B-ISDN terminal and the functions which must be carried out by the interworking function at the periphery of the ATM network.

ITU-T recommendation I.580 defines how B-ISDN networks (i.e. ATM networks) should interwork with *narrowband* ISDN networks. Since the capabilities of the interconnected networks are limited to the 64 kbit/s circuit switched capabilities of *narrowband* ISDN, the interface is clearly not of value to all possible *terminal equipment (B-TE)* connected to a B-ISDN. The interconnection only has value for narrowband ISDN terminal equipment (TE) or equivalent devices connected to the B-ISDN (B-NT2 type broadband terminal equipment).

Recommendation I.580 defines an interworking procedure in which the B-NT2 appears to be directly connected to the narrowband ISDN network, carrying information by means of AAL1 in an unchanged form to the narrowband ISDN network. The narrowband ISDN device connected to the B-NT2 uses normal narrowband ISDN signalling (in the ISDN *D-channel*) to setup or receive connections. This is carried transparently by the B-ISDN

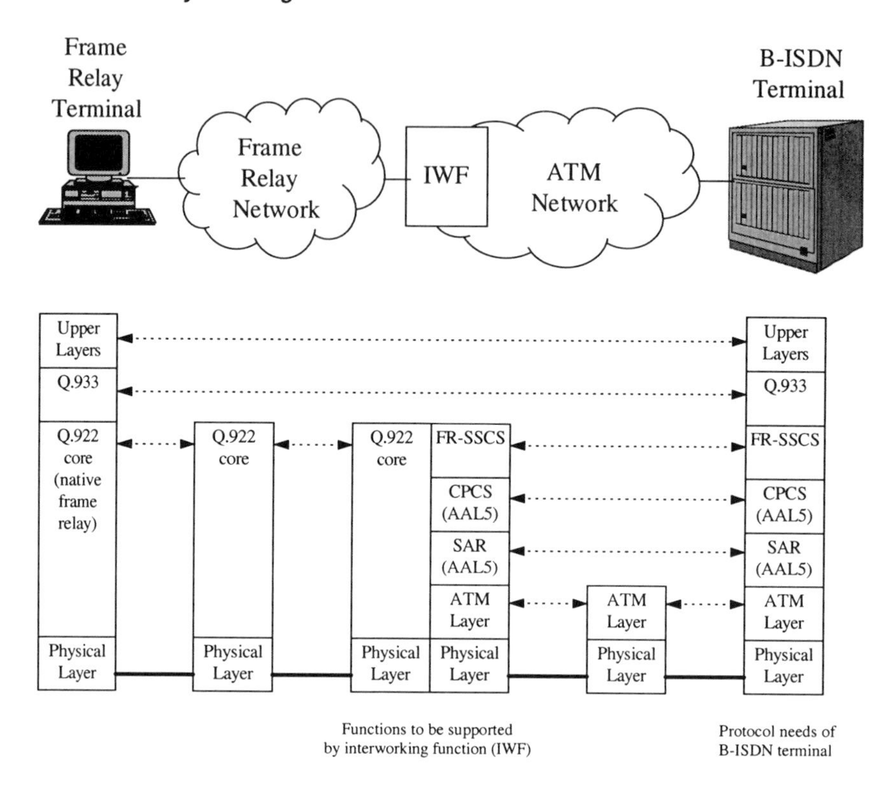

Figure 11.9 Interworking of frame relay and ATM networks according to ITU-T I.555

AAL1 to the *network adaptor (NA)* function (a specific name for the interworking function between ISDN and B-ISDN) which interprets the signalling information for onward switching of the connection in the narrowband ISDN network. The call setup is then furthered by means of normal ISDN network signalling through the *narrowband ISDN* (MTP/ISUP) or access signalling to the narrowband ISDN end user (Q.931/Q.921 (DSS1)) as appropriate (Figure 11.10(a)).

Once the connection is established (on the ISDN *B-channel*), the user information is carried transparently from one end device to the other (Figure 11.10(b)).

For calls originated in the narrowband ISDN network, the equivalent reverse procedure takes place. I.580 thus permits B-ISDN customers also to be narrowband ISDN users, and accessible by narrowband ISDN customers.

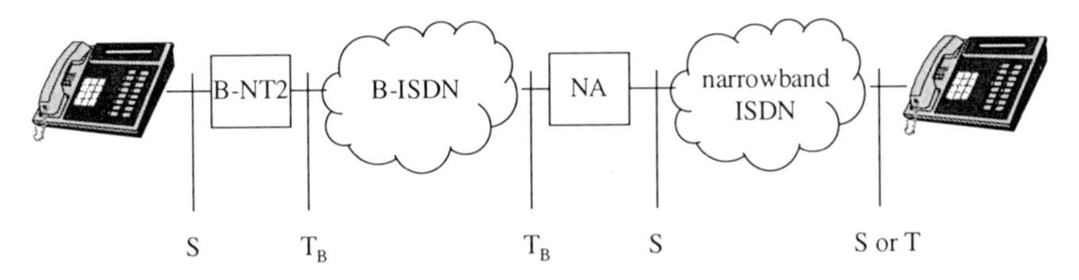

(a) Interworking of the ISDN connection setup signalling (D-channel)

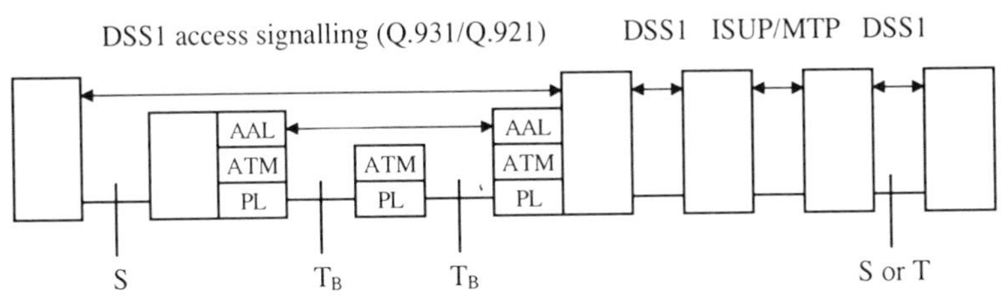

(b) Interworking of the ISDN 64 kbit/s user connection (B-channel)

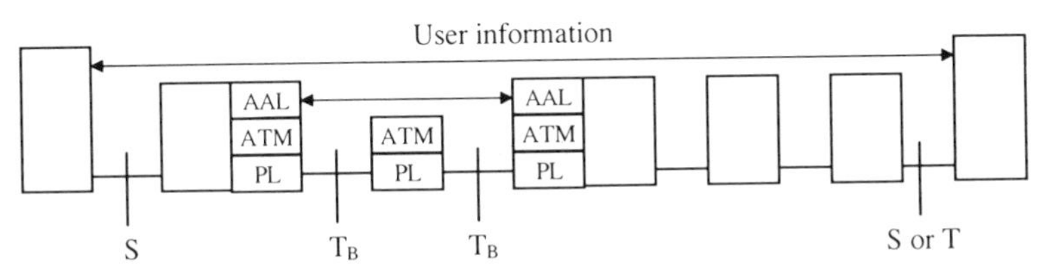

Note: ISDN provides for transparent circuit mode connections over which any user information or protocol may be transferred. The protocol boxes relating to 64 kbit/s ISDN are therefore, by convention, left blank.

Figure 11.10 B-ISDN interworking with 64kbit/s (narrowhead) ISDN according to ITU-TI.580

11.7 The Management, Control and Administration of ATM Networks

The performance management of ATM networks presents a major challenge to ATM network operators. For while the *statistical multiplexing* technique used in ATM affords, on the one hand, the potential to be highly efficient in the use of transmission bandwidth, it also brings the risk of encountering *Maxwell's Demon* – it might just happen that all the devices send cells at the same time (theoretically unlikely – but demonic if it happens). In this case unacceptable delays and variation in delays would occur.

While ITU-T recommendation I.356 ('B-ISDN ATM Layer cell transfer performance') lays the groundwork for a new science of ATM network traffic statistics, we have as yet little experience of real traffic and the interference effects of one traffic type upon another. We have no reliable statistical models of real traffic and only experience will fill this gap. In the meantime, if the network congests, the best advice is to add capacity at the bottleneck. It may lead to a less than optimal utilization of the network transmission resources as a whole, but it maintains the users' patience until our statistical modelling is better!

ATM network routing and its management also presents a major challenge to early ATM operators, since few standards are yet available. Before any connection can be established by an ATM network, some sort of decision must be made regarding the route the connection should take. (The same route is usually used to transport all the cells corresponding to a particular connection in order to minimize the risk of *cell misinsertion* or excessive *cell delay variation (CDV)*.) But which routes are available? And which of these is the best?

In the example of Figure 11.11, before a connection can be established between user A and user B, the network must first know which routes are available and must then be able to decide which route is most appropriate: that via exchanges CEFD, via CEGD, via CEGJD, via CHGD, via CHGJD or via CHJD.

The best route may depend upon the bitrate of the required new connection, upon the bandwidths of the various inter-node connections and upon current network loading. While the routes via GJD may not appear at first to make sense when a direct route GD is available, it may be that the required connection bitrate exceeds the bandwidth of the link GD but not that of route GJD.

Even in the relatively simple eight-node network of Figure 11.11 we have six potential routes for connections made between users connected to nodes C and D. It would take some time just to list all the other route permutations between all other individual node pairs. We then, as above, need to verify for

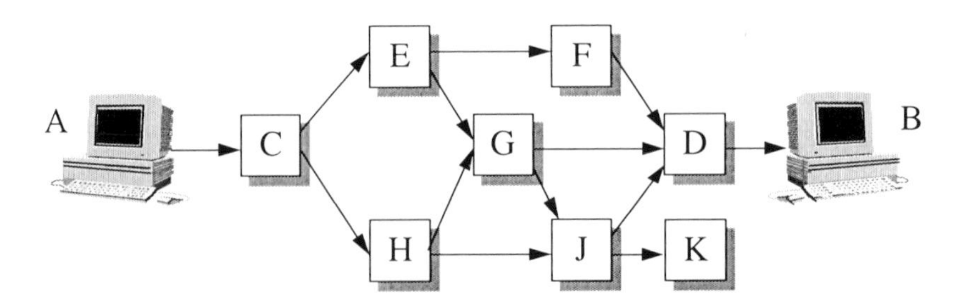

Figure 11.11 Possible routes through an ATM network

each route what the maximum bitrate of connections using the route may be. Finally, before we can decide whether new connections can be set up, we need to know how much bandwidth is being used by current connections.

As the number of nodes in the network increases, so the route permutations between any pair of nodes rapidly increases, and the task of deciding routes for new connections becomes exceptionally complicated, much more complicated than the human brain can find an optimal solution for. And anyway, if the network is continually being changed by the addition of new nodes and transmission links, the solution will only have a short lifetime. Much easier for the human network operator would be to rely on the network to determine its own best route for each new connection, taking into account current available transmission path permutations and traffic loading.

Apart from the ATM Forum *PNNI (private network–node interface)* specification, the standards are relatively silent on the matter of ATM network routing, and the various switch equipment manufacturers have elected to use different methodologies. Some manufacturers require the human network manager to input allowed route permutations between each pair of endpoints and list the routes in preferred choice order. Others also use fixed routing tables, but then employ a *metric* to adjust the route choice according to prevailing network loading. The advantage of fixed routing tables is the tighter administrative control over route choice by user devices. The disadvantage is the huge amount of manual effort needed to keep the routing tables up-to-date with continual changes in network topology.

In the most sophisticated networks (e.g. IBM's *broadband network service (BBNS)* – the network architecture behind its *NWays* ATM switch), a modified *spanning tree protocol (STP)* method is used to determine the best current route. It does this by *polling* the network for the best path available at that instant. A spanning tree protocol applied to our example of Figure 11.11 might work as described below.

Node C broadcasts a message to all neighbouring nodes: 'I need a connection of bitrate X to address B, can you help?'. Nodes E and H receive the poll, and broadcast their own similar messages, respectively to nodes G and F and to nodes G and J. Node C is excluded from the broadcast, as it is the source of the enquiry. The polling continues – as nodes F, G and J enquire of node D and node K . Node D's response is positive: 'I own this port address'. The message is relayed via the intermediate nodes F, G and J, and in turn nodes E and H are also able to return positive replies, appending their own information about route conditions: 'maximum bandwidth currently available is . . .' and 'total number of hops to the destination is . . .' etc. The information is extended at each point of its return to node C. Finally C decides from the two responses (from E and H) which is best by comparing the *attributes* and *metrics* of the alternatives, and sets up the connection.

Being on a 'blind alley', node K returns a negative response to node J, and the polling is not furthered by node K. How does node K recognize that it is a blind alley? Simply because it has no possibility for further polling but does not recognize the requested address.

The spanning tree method of finding routes through a network is largely a product of the *Internet*, where such a methodology is one of only a few ways of finding a path through an impossibly complex and ever-growing worldwide labyrinth of computer or router networks owned by many thousand different companies. The spanning tree methods available within the Internet protocols are, however, more sophisticated than that of our example. It would, after all, be ridiculous to poll every Internet node in the world every time any Internet user sent a message or set up a connection.

The advantage of using a spanning tree protocol is that the routing tables and algorithms for the network are maintained by the network automatically, without human intervention. In addition, they are able to adapt connection paths to take into account current traffic loading patterns and system outages or failures. A spanning tree method of route control (*routing*) can thus contribute to the optimal dynamic loading of the network.

The disadvantage is the heavy loading inflicted upon the network in administering itself. This adds message overhead traffic to the transmission links and to the switch control processors. In addition, it adds delay to the process of setting-up connections.

It is too early to say which method of route determination and management will turn out to be the best, but it is wise to ensure that your network equipment manufacturer is aware of the problems.

Assuming our network manager copes with the hurdle of creating reliable planning models for predicting the loads on the network, and also with the difficulty of managing routes through the network for individual connections, then he is likely to satisfy his customer's main quality demands. His final concern will be to ensure that the customer pays for his usage. This will rely on the ability of his network to generate meaningful and adequate *accounting records* (i.e. messages generated by the network containing information about the connection endpoints, connection start time and duration, bitrate, volume of data carried, etc.). This information needs to be processed by a billing system to create an invoice or internal company transfer charges bill for the end-user.

The production of the accounting records by the ATM network is unlikely to be a problem, providing we can decide what information we need to record about each connection. Herein lies the greater dilemma. What is the best basis for tariffing a connection made across an ATM network? Should we account and tariff as for a telephone call, simply based on bitrate and connection duration? Or like a packet switched data connection, where connection duration is of secondary importance, and instead a charge is made for the total number of bytes sent, independent of how long it took?

My guess is that ATM network operators will need every possible accounting record variant. It seems to me that it makes most sense to account telephone connection users of ATM networks according to the tariff structure they are used to – a circuit charge per minute. Meanwhile the data users will expect the charge per byte transmitted. Illogical, you might say! And maybe it seems so, if you then discover that the charge per ATM cell is different for the telephone call than for the data connection. This might seem unfair, but why should it be unfair to charge the customer exactly what he expects to pay?

And the advice when buying new ATM network switching equipment? Make sure the accounting records are very comprehensive.

11.8 Access Line Infrastructure

It is a fact that less than half of the telecommunications transmission capacity available to public telecommunications network operators is actually in use. On the one hand, this pool of unused resources provides some scope for the rapid expansion of a broadband network services market – the achievement of the *information highway* about which our politicians philosophize. On the other hand, this capacity is not geographically evenly spread. In different localities the available supply varies greatly.

The greatest problem will be in the 'last mile' – the connection of end customers to the network. In the same way that highways are no good without connecting roads, so the user's connection circuit is the bottleneck of most telecommunications networks. Thus while huge overcapacity exists in long distance networks, customer lines remain inadequately provisioned.

In some telecommunications markets there are already numerous companies actively installing modern cable television networks. These will greatly improve the available supply, particularly for individuals and companies located in conurbations.

In remote and less densely populated areas, the costs of laying new cables will preclude the availability of the fibre connection ideal for ATM network access. For these areas, potential network access solutions include ATM at 25 Mbit/s, 13 Mbit/s or even 2 Mbit/s using unshielded telephone wire copper pairs. Meanwhile, there is also great scope for ATM via radio. Special techniques are likely to be necessary to accommodate ATM in the burst error-prone world of radio. On the other hand, radio is an ideal match for broadcast applications and sharing network capacity between many bursty traffic sources.

12
The First Uses of ATM

The outstanding qualities of ATM when compared with preceding technologies are its wide range of supported bitrates, and its potential for efficient integration of voice, data and other signal types. These qualities (once fully developed) make it an ideal backbone technology, with universal application. It will be built into all sorts of communicating devices and be the basis of many networks. Even so, the early application fields will be in three niche markets – wide area corporate backbone networks, corporate campus networks, and public switched broadband. In this chapter we discuss why, and predict how future applications will evolve.

12.1 ATM as a Corporate Backbone Technology

As early as the 1980s, 'fast packet switching' techniques (the predecessors of the ATM cell relay technique) were establishing themselves within the market for *wide-area* corporate backbone networks. In particular the *Stratacom* company established a 24-byte cell switching technique supported by its *IPX* product, and has won many major corporations as its customers. The Stratacom product has become one of the leading *frame relay* network switch products and a strong contender to the classical TDM (time division multiplex) technology of corporate transmission networks (for breaking down high-speed leased lines into lower bitrate user channels). The product is well suited to data networks and for the integration into these networks of point-to-point telephone or voice connections. Stratacom was one of the first suppliers to offer *silence suppression* of voice connections.

The Stratacom product has been successful for two main reasons. First, because of its exceptional support for the frame relay service. This extended the scope of data networking to 2 Mbit/s bitrates, where 64 kbit/s had been the earlier practical limit for X.25–based packet networks. The second Stratacom success was the efficient integration of point-to-point connections for internal corporate voice networks carried within a corporate data network.

The fact that the NNI used between Stratacom switches was a proprietary (i.e. Stratacom-specific) interface was not a hindrance to the Stratacom success. Indeed, it was so much of a success that the 24-byte cell switching technique used became the seed from which today's 53-byte ATM cell switching has developed. The product sold because it supported the user interfaces expected of corporate networks, and because it made those backbone networks more efficient than had hitherto been possible. For similar reasons, ATM switches can be expected to be popular amongst corporate network buyers.

ATM-compliant devices will provide direct competition for the established Stratacom IPX product in the supply of corporate backbone networks. Thus, as Figure 12.1 illustrates, an ATM-based product with AAL functionality is like the original Stratacom offering in its ability to offer connections between frame relay devices or point-to-point (CBR) connections for point-to-point voice or telephone needs. The difference is that the ATM network is based on standardized interfaces, and therefore offers the potential for multi-vendor supply and extension to the broadband bitrates of ATM.

One of the earliest broadband network uses of wide area ATM corporate (and also public) networks will be for the interconnection of campus LAN networks. No longer will the data and software sharing capabilities of the headquarters network be limited to the headquarters building! ATM can be used to convince even remote users that they are locally connected, since the

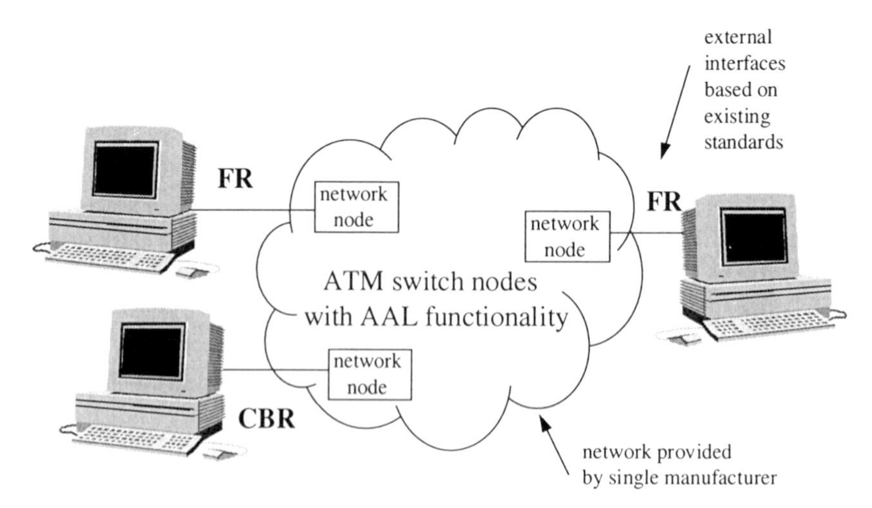

Figure 12.1 A *wide area* corporate backbone network based upon early ATM

34 Mbit/s, 45 Mbit/s and 155 Mbit/s speeds of ATM are even faster than the typical LAN speeds of 10 Mbit/s (Ethernet) and 16 Mbit/s (Token Ring).

12.2 The Appearance of *Hybrid* Networks

ATM equipment manufacturers who take on the market of providing equipment for corporate backbone networks will have not only Stratacom to compete with. The market is also contended by a number of longer established companies, who have historically provided TDM based multiplexor products to this market. These companies include Timeplex, Newbridge, GDC and NET.

The marketing trend amongst some of the historic TDM suppliers has been to point out the inherent difficulties faced by ATM networks in carrying *clear channel* (or *constant bitrate, CBR*) connections (such as tie-line telephone connections between PBXs on different company sites). The difficulties are associated with the guarantee of a constant bitrate and propagation delay (cell delay variation), and the potential bandwidth overhead caused by the cell header. Some of the TDM manufacturers have therefore set about the development of *hybrid* products which use a mix of the TDM and ATM techniques within the network.

Their technical goal is to achieve a product offering the best of both ATM and TDM – a network with the ability to decide which transmission technique is best suited to each individual connection in order to achieve a high level of performance and an optimum usage of the trunk bandwidth. These devices segregate the trunk bandwidth into two separate pipes of variable size, one used in TDM mode, the other in ATM or frame relay mode (Figure 12.2). The sizes of the two pipes are adjusted dynamically over time in order to cope with

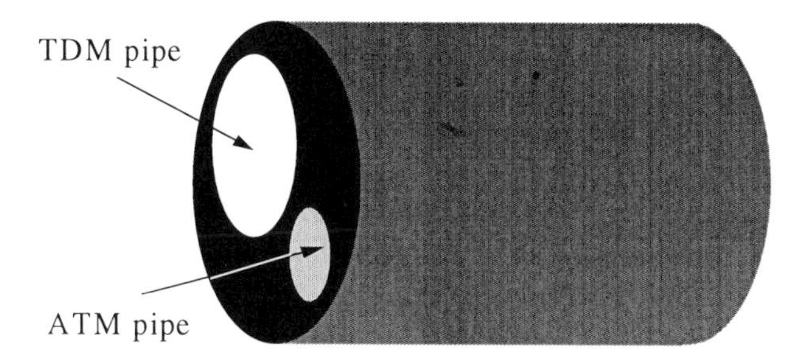

Figure 12.2 A *hybrid* transmission technique separating bandwidth between TDM and ATM

instantaneous demand. This is done by reallocating bandwidth from TDM to ATM/frame relay or vice-versa.

The ATM device manufacturers have, meanwhile, also noticed that the rigid sized 53-octet cell is not ideally suited to carrying all the different information types. In particular, data frames (in X.25 and frame relay) are typically 256 or 1024 octet (byte) length. Such frames therefore occupy respectively 6 and 21 cell payloads (of 48 octets). The maximum percentage of useful information carried is thus $256/(6 \times 53)$ and $1024/(21 \times 53)$ respectively, or 81 per cent and 92 per cent. These figures compare very unfavourably with the frame relay maximum network efficiency exceeding 98 per cent. So what is the solution? Hybrid technology. Amongst others, both Northern Telecom (Nortel) and IBM, for example, in their *Magellan Passport* and *NWays* products respectively, have chosen to implement trunk transport protocols which are a mixture of a cell-oriented switching and a frame relaying technique, where the backbone trunk speed is 34 Mbit/s or lower. The technique is referred to as *frame/cell switching*.

The frame relay/ATM hybrid (*frame/cell switching*) seeks to improve the trunk bandwidth utilization by allowing for the transmission of large data frames as frames rather than subjecting them to segmentation into ATM cells. The frames are thus sent unsegmented on the trunk. However, should a higher priority cell (e.g. a constant bitrate cell) appear in the send queue before the frame transmission is complete, then the frame is interrupted and the cell is sent. The frame transmission then continues (Figure 12.3).

Maybe the various different approaches to the development of hybrid products will lead to a further strengthening of the ATM technique.

12.3 ATM Multiplexors

One of the earliest types of 'native' ATM end-user equipment will be various types of multiplexors and ATM 'PADs', which serve the simple function of combining multiple point-to-point connections of existing types (e.g. constant

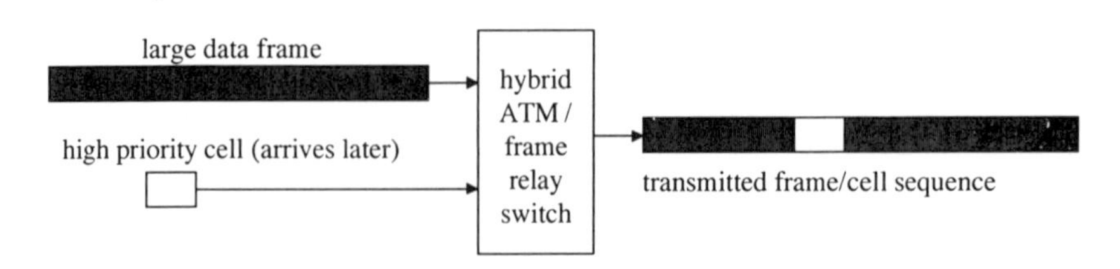

Figure 12.3 A hybrid ATM / Frame relaying switch device

bitrate,CBR, or frame relay PVC) across a single line (Figure 12.4). In conjunction with ATM crossconnect devices, ATM multiplexors used as ATM network end equipment provide an alternative to an existing TDM network. This is the type of ATM network offered by some of the earliest public ATM operators (including the public telephone companies).

Such networks are technologically advanced compared to the TDM networks they replace, but do not offer significant extra benefits or potential uses other than the range of bandwidths they make available and the possibility of being a cheaper alternative for a high speed leaseline.

12.4 The ATM Data eXchange Interface (DXI) and HSSI (High Speed Serial Interface)

For customers or users who would like to start on the migration path to ATM but do not yet warrant the high data speeds which it offers in its native UNI form, the *Data eXchange Interface (DXI)* offers a first step.

DXI is a DTE/DCE (data terminal equipment / data circuit terminating equipment) interface. It provides the ability of a pre-ATM device (such as a router) to cooperate with an ATM DCE (such as a *digital service unit (DSU)* or network/line terminating device) to use a native ATM UNI, requiring only relatively minor software enhancements. The DXI protocol may be used in conjunction with any of the physical interfaces V.35, X.21(RS449) or the 58 Mbit/s *HSSI (high speed serial interface)*. The advantage for the end device is the higher speed thereby made possible without major hardware redesign.

The ATM DXI (Figure 12.5) is a full ATM interface, but caters for

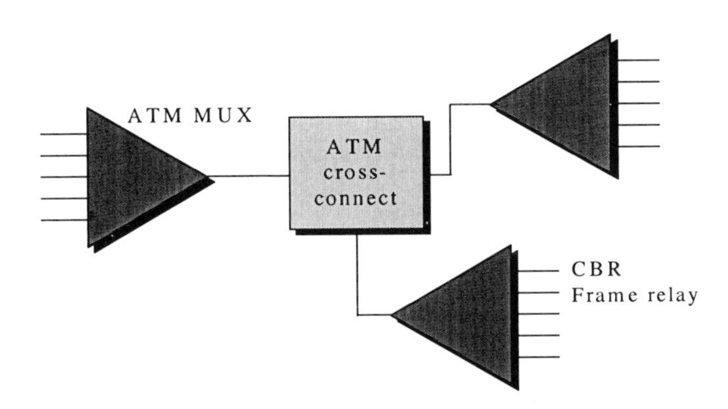

Figure 12.4 An early ATM network comprising ATM multiplexors and crossconnects

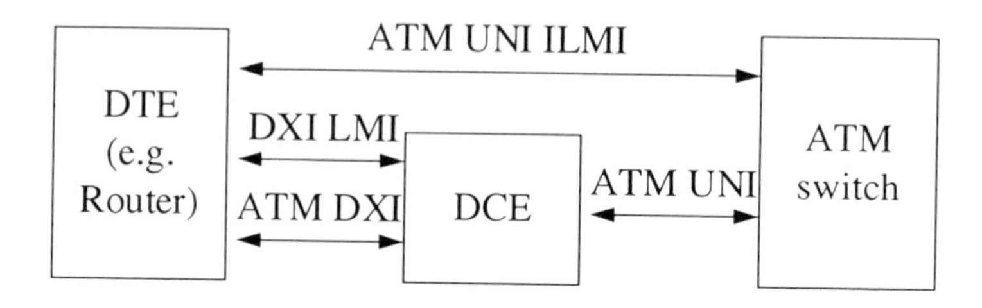

Figure 12.5 The ATM Data eXchange Interface (DXI)

point-to-point data connections of bitrates in the range 64 kbit/s to 58 Mbit/s. The DXI is likely to be offered by ATM DSUs, ATM multiplexors and ATM PADs. It is similar to the SMDS DXI (switched multimegabit data service DXI), but not identical, and thus likely to make the SMDS DXI obsolete.

Management of the DCE device and the DXI configuration takes place using a *local management interface (LMI)* between the DTE and the DCE. The management messages associated with the control of the UNI between the DCE and the ATM network switch take place via the *ILMI (Interim local management interface)*, though the difference between LMI and ILMI may not be clear to the DTE. LMI and ILMI are both similar to the *simple network management protocol (SNMP)* which is commonly used for the management of data networks.

The ATM DXI can be operated in any of three modes, 1a, 1b and 2, the characteristics of which are summarized in Table 12.1. The main difference between the modes is in the number of VCCs which can be supported and in the maximum size of the DTE service data unit (SDU).

The DTE must be capable of supporting either AAL3/4 or AAL5, into which the service data unit (SDU – i.e. the data frame) must be packed. The frame check sequence used in the 1a and 1b modes is the same sequence used by HDLC (high level datalink control). This brings the possibility of adaptation of existing HDLC or frame relay devices for support of the ATM

Table 12.1 The various ATM DXI modes

Characteristic	Mode 1a	Mode 1b	Mode 2
Maximum number of VCCs	1023	1023	16777215
AALs supported	AAL5 only	AAL3/4 or AAL5	AAL3/4 or AAL5
Maximum SDU size in octets (bytes)	9232	9232 (9224 for AAL3/4)	65535
Frame check sequence (FCS) length in bits	16	16	32

DXI mode 1a or 1b. The most likely physical interfaces are then V.35 or X.21 (RS449) as commonly used in the 64 kbit/s to 2 Mbit/s range.

DXI Mode 2 uses a longer field for the connection addressing (24 bits as opposed to 10) and therefore brings with it more scope and the need for higher bitrates. The mode 2 interface is most likely to be offered by newer types of devices using the HSSI at 58 Mbit/s. A potential use of this interface, for example, is for the support of host computer channel extension.

12.5 Campus ATM Networks – the Next Generation Local Area Network (LAN)

Campus networks will present an early market for ATM equipment. On large corporation campuses there is already the need for high bandwidths to resolve the problems arising in the *local area* networking of personal computers, printers, file servers and the like. In addition, on many campuses the fibre cabling needed for ATM is already available in plentiful supply. Unlike *wide area* ATM networks, local area and campus ATM networks will not be held back by shortage of line capacity or the cost of the lines. On the other hand, the efficiency of the ATM technique in integrating different information types is unimportant, since lines on campus are relatively cheap anyway. So, you might say, why bother with ATM? There are two main reasons:

1. The high speed bitrates of ATM could be the key to resolving overload problems in *LANs* (*local area networks* of personal computers). The 155 Mbit/s rate of ATM could provide the basis for a much faster and more capable type of LAN than today's Ethernet (10 Mbit/s) or Token Ring (4 Mbit/s or 16 Mbit/s).
2. The fact that ATM appears to offer a real potential to become the universal networking technology.

Figure 12.6 illustrates the typical hub and router devices used today in a *local area network*. The hub is the wiring centre to which each of the end-user devices is connected by direct wiring, usually laid in a 'star' fashion. The hub provides for the correct physical connection of the devices making up the LAN, creating either a *logical bus* (as in the case of an Ethernet LAN – Figure 12.7(a)) or a *logical ring* (the case of a Token Ring LAN – Figure 12.7(b)).

Most of the Router and Hub manufacturers are developing or have already developed ATM cards and software for their equipment, so that in principle any or all of the connections of Figure 12.6 could be equipped with ATM technology.

First affected is likely to be the router–router connection. If both routers are

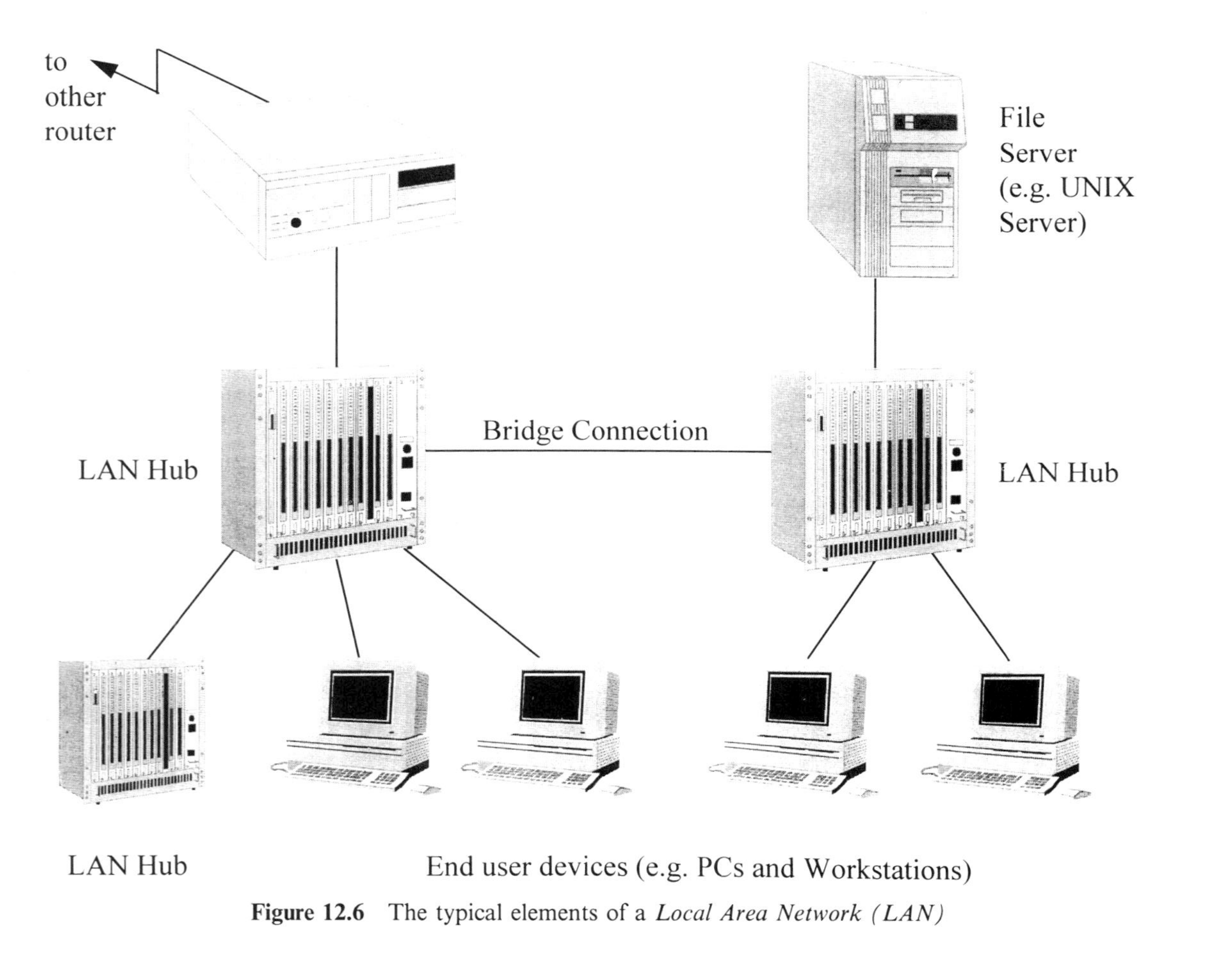

Figure 12.6 The typical elements of a *Local Area Network (LAN)*

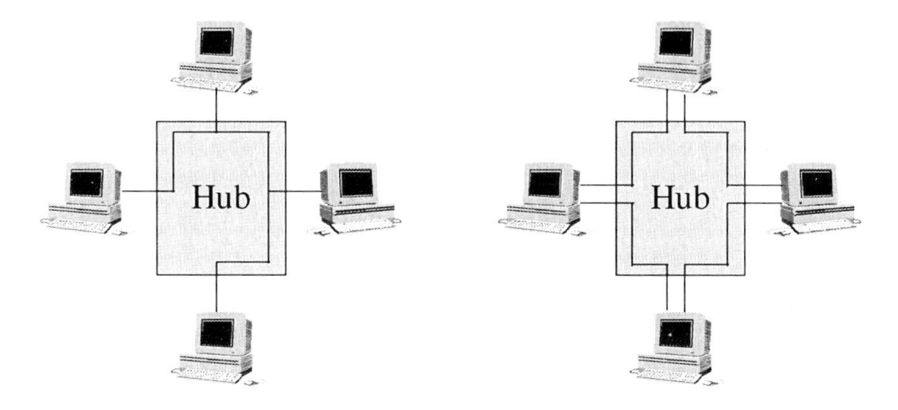

Figure 12.7 Different logical LAN connections achieved through hub wiring

on campus, this connection, if already existing, is likely to be 100 Mbit/s FDDI or SMDS, running over a campus glassfibre backbone. This can be immediately reconfigured using 'native' ATM cards in the routers.

Where the routers are geographically remote from one another the connection is likely to be frame relay, and relatively low speed (in the range 64 kbit/s to 2 Mbit/s). This is one of the major bottlenecks of existing networks of interconnected LANs. Over the short term, the frame relay service of a public ATM (the *frame relay user network interface (FUNI)*) could be used. Over the medium term, the use of native ATM protocols over a corporate ATM network or a public ATM network will solve the problem.

As an option, the router-to-router connections could use the ATM DXI (HSSI) interface to provide higher speed point-to-point connections. This has the benefit for an existing router network operator that ATM can be introduced without disturbing the existing network topology or operating software. As the router network evolves the connection could be evolved to a full ATM UNI and thus towards the new specific ATM router protocols (e.g. the IETF RFC 1577 standard describes how the Internet protocol (IP) can be carried by an ATM switched virtual connection).

The UNIX server is likely to be next converted to ATM. This will provide for much faster file transfers and prepare the way for *multimedia* applications. A number of major UNIX hardware providers are already active (SUN, Hewlett Packard, IBM, DEC). At the same time as the server is upgraded, certain end-user workstations are likely to be creating the need for fibre to the end-user, and finally, fibre to every desktop is inevitable. At this stage, the hubs also need to have ATM cards, and historical Ethernet and Token Ring LAN operating software needs to be upgraded to new versions which are ATM compatible. The *LAN emulation (LANE)* standards define the

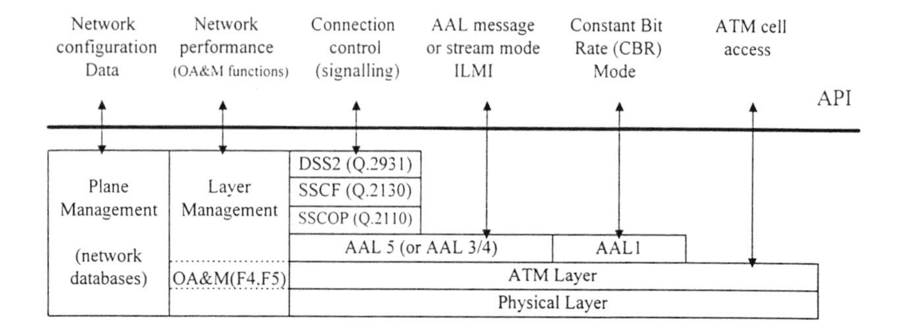

Figure 12.8 ATM application programming interfaces (APIs)

various functions which need to be carried out by the various devices, and the protocols to be used. In particular, the *LUNI (LAN user–network interface)* defines ATM-based protocols to replace today's Ethernet and Token Ring ones.

On his desk of tomorrow, the business executive will expect to have a workstation far more powerful than his current PC. Many manufacturers are already conceiving a new type of 'one-per-desk' device, capable of performing the functions of what today requires many devices. Via multimedia applications and high speed communications these devices will be able to supply information and control capabilities to one's individual needs. And the price is likely to be in the same range as a PC of today. Compaq, Apple and Xerox, amongst other companies, are working on such devices. Maybe he won't have to wait long, but in the meantime our business executive will have to be satisfied with an ATM card which he can slot into his PC. They already cost less than $1000!

And what, might you ask, will happen to the office telephone system – the PBX? ATM switch capabilities will be added to most major PBXs, and in the longer term we can expect the entire switching in PBXs to be carried out using ATM. The use of ATM will free telephone and video switching from the constraints of 64 kbit/s circuit switching, allowing lower or higher rates as the user and his quality expectations demand. In addition, an ATM PBX will allow the internal wiring of buildings and campuses to be ATM-based, so that even the telephone extensions can be connected to the PBX using ATM. The short term difficulty for the PBX manufacturers is how to retain the benefits of the rich software functionality available in modern PBXs while converting the switch matrix to ATM. Corporate users have, meanwhile, come to rely on *ring back when free, conference facility, voicemail* and *call diversion on busy.*

Amongst other telephone switchboard manufacturers, Siemens has announced an ATM capability for its HICOM PBX family.

12.6 ATM Application Programming Interfaces

ATM *Application Programming Interfaces (APIs)* will define standard controls and procedures for using the various capabilities of an ATM network (figure 9). A number are under study by the ATM forum, and at ITU-T under the banner of a *programming communication interface (PCI)*.

The definition of standard ATM APIs will ensure that computer software applications are able to make the best use of ATM network capabilities without requiring extensive software amendment to cope with the ATM. Thus by using a standard signalling API, a device or software can set up a connection, and then by using the AAL message mode, CBR mode or cell access API can send information easily across the connection. The application software provider can thus adapt programs to ATM simply by amendment of existing APIs (written, for example in IBM's *Netbios,* Microsoft's *NDIS,* Hayes *AT* or Novell Netware's *IPX*).

A standardized ATM API set will go a long way to removing any remaining resistance to the migration to ATM. Furthermore, the extra capabilities of ATM (point-to-multipoint, greater choice of connection types and qualities) will create new application possibilities.

12.7 ATM Applications and Multimedia

Once the base development and investment has been made in ATM, there will be scope for the appearance of new applications making use of the particular capabilities of ATM. These will be in the realms of *broadband* and *multimedia.* The earliest such applications are likely to be for workgroups, enabling individuals in different geographical locations to work together, as if they were in the same room. Initially the focus will be on video meeting type facilities. As time progresses, *groupware* software (similar to *Lotus Notes*) will allow groups of individuals to create documents and databases together. Maybe scientific experiments can be carried out under the control of remote teams of staff! There is also considerable development being applied to remote education and medicine – even remote surgery! A new era is emerging, and a complete new vocabulary to go with it (ATM, tele-education, tele-medicine, tele-working!).

You may, like me, be a little sceptical about the first *multimedia* applications which are appearing and whether we really need them, but within a few years we may wonder how we managed without ATM and multimedia.

Appendix 1

The *Synchronous Digital Hierarchy (SDH)* and *Synchronous Optical Network (SONET)*

The *Synchronous Digital Hierarchy (SDH)*

The *synchronous digital hierarchy* (*SDH*) was developed based upon its North American forerunner *SONET (synchronous optical network)*. SDH is the most modern type of transmission technology and, as its name suggests, is based upon a synchronous multiplexing technology. The fact that SDH is synchronous adds greatly to the efficiency of the transmission network and makes the network much easier to manage.

Historically, digital telephone networks, modern data networks and the transmission infrastructures serving them have been based on a technology called *PDH (plesiochronous digital hierarchy)*. Three distinct types of PDH structure exist as illustrated in Figure A1.1. They share three common attributes:

- they are all based on the needs of telephone networks – i.e. offering integral multiples of 64 kbit/s channels;
- they require multiple multiplexing stages to reach the higher bitrates, and were therefore difficult to manage and relatively expensive to operate;
- they are basically incompatible with one another.

Each individual transmission line within a PDH network runs *plesiochronously*. This means that it runs on a clock speed which is nominally identical to all the other line systems in the same operator's network but is not locked *synchronously* in step. This results in certain practical problems. Over a

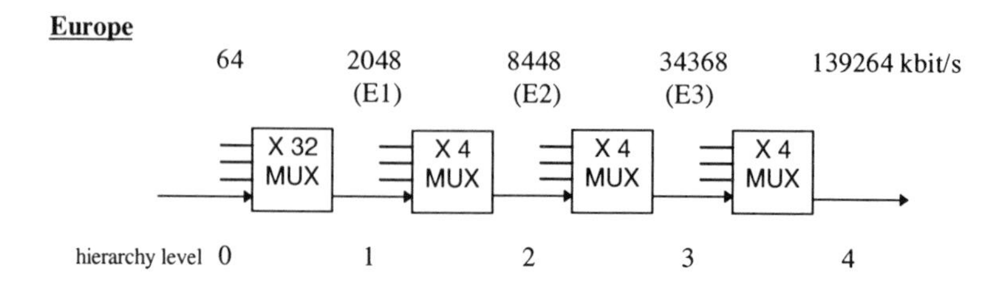

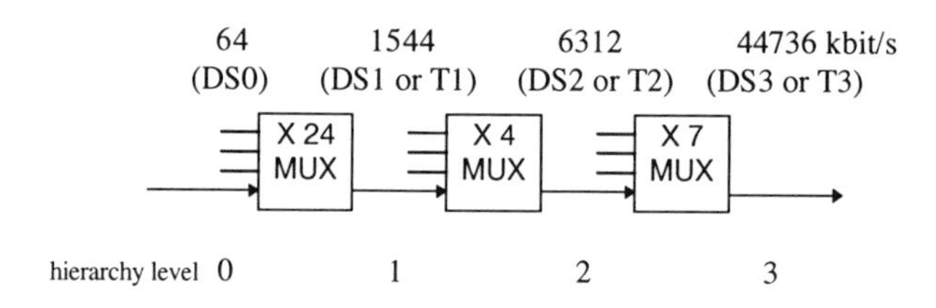

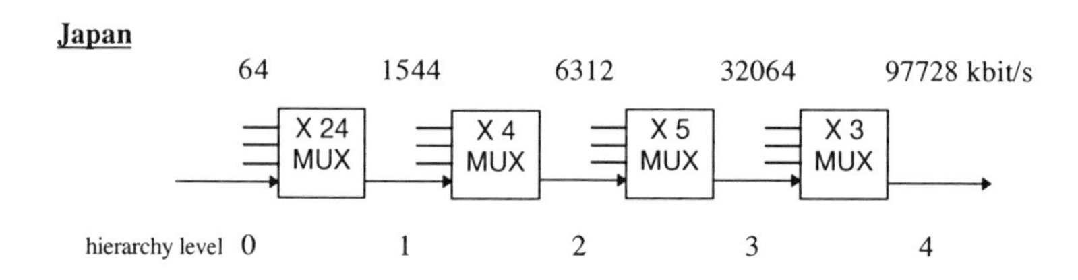

Figure A1.1 The various plesiochronous multiplexing hierarchies (ITU-T/G.571)

relatively long period of time (say one day) one line system may deliver two or three bits more or less than another. If the system running slightly faster is delivering bits for the second (slightly slower) system then a problem arises with the accumulating extra bits.

Eventually, the number of accumulated bits become too great for the storage available for them, and some must be thrown away. The occurrence is termed *slip*. To keep this problem in hand, *framing* and *stuffing* bits are added within the normal multiplexing process, and are used to compensate. These bits help the two end systems to communicate with one another, slowing up or slowing down as necessary to keep better in step with one another. The extra framing bits account for the difference, for example, between 4 × 2048 (E1 bitrate) = 8192 kbit/s and the actual E2 bitrate (8448 kbit/s – see Figure A1.1).

Extra framing bits are added at each stage of the PDH multiplexing process.

Unfortunately this means that the efficiency of the higher order line systems (e.g. 139 264 kbit/s – usually termed 140 Mbit/s systems) is relatively low (91 per cent). More critically still, the framing bits added at each stage make it very difficult to break out a single 2 Mbit/s *tributary* from a 140 Mbit/s line system without complete demultiplexing. This makes PDH networks expensive, rather inflexible and difficult to manage.

SDH, in contrast to PDH, requires the synchronization of all the links within a network. It uses a multiplexing technique which has been specifically designed to allow for the *drop and insert* of the individual *tributaries* within a high-speed bitrate. Thus, for example, a single *drop and insert multiplexor* is required to break out a single 2 Mbit/s *tributary* from an *STM-1 (synchronous transport module)* of 155 520 kbit/s.

As is shown in Figure A1.2, the *containers* of the *Synchronous Digital Hierarchy* have been designed to correspond to the bitrates of the various PDH hierarchies. These *containers* are multiplexed together by means of *virtual containers* (*vcs*, but not to be confused with atm's *virtual channels*), *tributary units (TU), tributary unit groups (TUG), administrative units (AU)* and finally, *administrative unit groups (AUG)* into *synchronous transport modules (STM)*.

The basic building block of the SDH hierarchy is the *administrative unit group (AUG)*. An AUG comprises one AU-4 or three AU-3s. The AU-4 is the simplest form of AUG, and for this reason we use it to explain the various terminology of SDH (containers, virtual containers, mapping, aligning, tributary units, multiplexing, tributary unit groups).

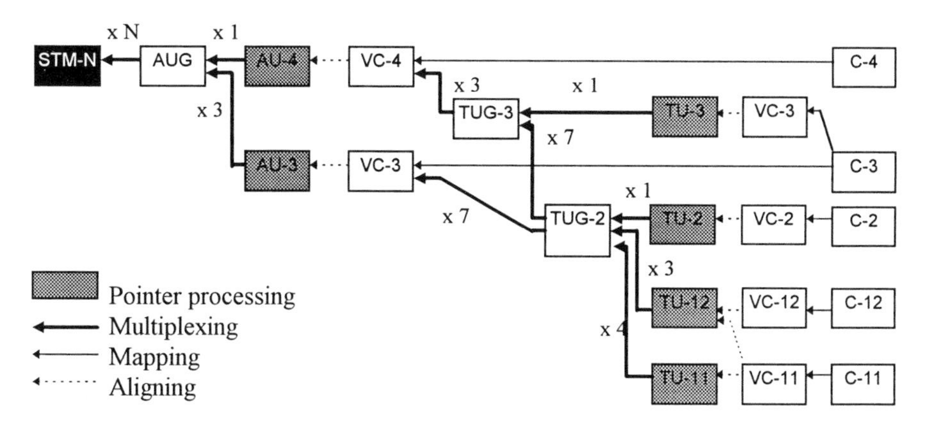

Figure A1.2 Synchronous Digital Hierarchy (SDH) multiplexing structure (ITU-T/G.709)

The *container* comprises sufficient bits to carry a full *frame* (i.e. one cycle) of user information. In the case of container 4 (C-4) this is a field of 260 × 9 bytes (i.e. 18720 bits). In common with PDH, the *frame repetition rate* (i.e. number of cycles per second) is 8000 Hz. Thus a C4-container can carry a maximum user throughput rate (*information payload*) of 149.76 Mbit/s (18.720 × 8000). This can be used either as a raw bandwidth or, say, to transport a PDH link of 139.264 Mbit/s.

To the container is added a *path overhead (POH)* of 9 bytes (72 bits). This makes a *virtual container (VC)*. The process of adding the POH is called *mapping*. The POH information is communicated between the point of *assembly* (i.e. entry to the SDH network), to enable the management of the SDH system and the monitoring of its performance.

The *virtual container* is *aligned* within an *administrative unit (AU)* (this is the key to synchronization). Any spare bits within the AU are filled with a defined filler pattern called *fixed stuff*. In addition, a *pointer* field of 9 bytes (72 bits) is added. The *pointers* (3 bytes for each VC – up to three) indicate the exact position of the virtual container(s) within the AU frame. Thus an AU-4 contains one 3-byte pointer indicating the position of the VC-4. The remaining 6 bytes of pointers are filled with an idle pattern. One AU-4 (or three AU-3s) are *multiplexed* to form an AUG.

To a single AUG is added 9 × 8 bytes (576 bits) of *section overhead (SOH)*. This makes a single *STM-1 frame* (of 19440 bits). The SOH is added to provide for *block framing*, maintenance and performance information carried on a transmission line *section* basis. The SOH is split into 3 bytes of *RSOH (regenerator section overhead)* and 5 bytes of *MSOH (multiplex section overhead)*. The RSOH is carried between, and interpreted by, SDH line system *regenerators* (devices appearing in the line to *regenerate* laser light or other signal, thereby avoiding signal degeneration). The MSOH is carried between, and interpreted by, the devices assembling and disassembling the AUGs. The MOH ensures integrity of the AUG.

Since the *frame repetition rate* of an STM-1 frame is 8000 Hz, the total line speed is 155.52 Mbit/s (19440 × 8000). Alternatively, power of 4 (1, 4, 16, etc.) multiples of AUGs may be multiplexed together with a proportionately increased section overhead, to make larger STM frames. Thus an STM-4 frame (4 AUGs) has a frame size of 77760 bits, and a line rate of 622.08 Mbit/s. An STM-16 frame (16 AUGs) has a frame size of 311040 bits, and a line rate of 2488.32 Mbit/s.

Tributary unit groups (TUGs) and *tributary units (TUs)* provide for further breakdown of the VC-4 or VC-3 payload into lower speed tributaries, suitable for carriage of today's T1, T3, E1 or E3 line rates (1.544 Mbit/s, 44.736 Mbit/s, 2.048 Mbit/s or 34.368 Mbit/s).

Figure A1.3 shows the gradual build-up of a C-4 container into an STM-1 frame. The diagram conforms with the conventional diagrammatic represen-

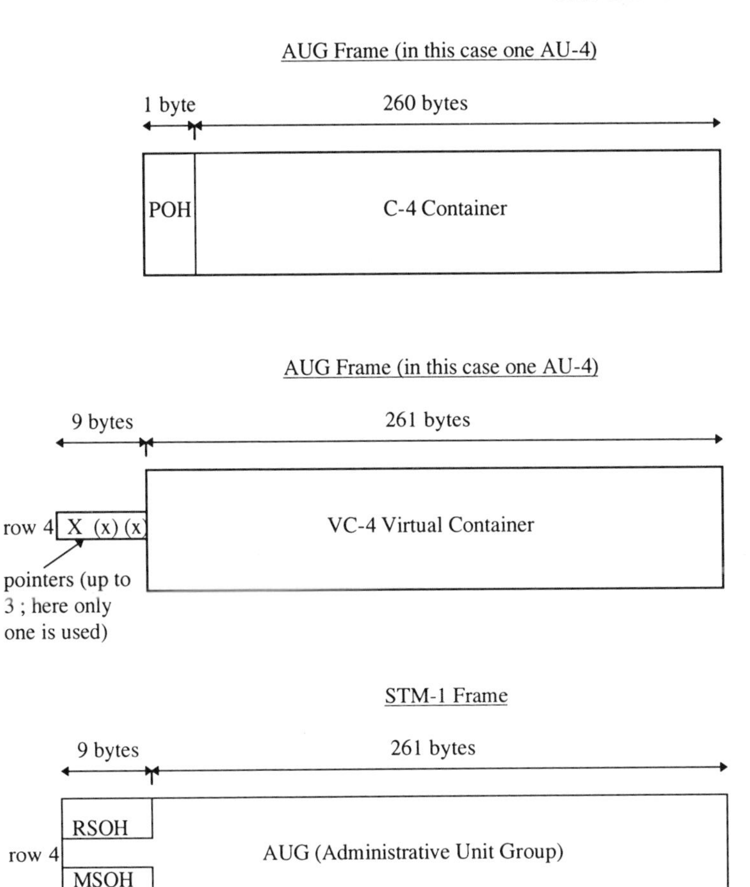

Figure A1.3 Structure of an STM-1 frame

tation of the STM-1 frame as a matrix of 270 columns by 9 rows of bytes. The transmission of bytes, as defined by ITU-T standards, is starting at the top left-hand corner, working along each row from left to right in turn, from top to bottom row.

The structure of an AUG comprising three AU-3s is similar except that the area used in Figure A1.3 for VC-4 is instead broken into three separate areas of 87 columns, each area carrying one VC-3. In this case all three pointers are required to indicate the start positions within the frame of the three separate VCs. The various other TU and VC formats follow similar patterns to the

AUs and VCs presented (TUs also include pointers like AUs). The various patterns are not explained in detail here. Instead, Table A1.1 simply presents the various container rates available within SDH.

Using a C-4 container at its full capacity (i.e. 149.76 Mbit/s) we achieve a system efficiency using SDH of 96% (c.f. 91% with PDH). The C-4 container may be used directly for carriage of ATM, and will be one of the standard speeds at which ATM will be used. ATM cells (of 53 bytes or *octets*) do not fit an integral number of times into the C-4 frame (2340 bytes), but this is not important. The SDH standards require only that the ATM *octets* are aligned with the bytes of the SDH container. Individual ATM cells can be split between container frames when necessary.

Compared with PDH networks, SDH networks are more efficient and easier to administrate (due to the availability of *drop and insert* multiplexors). But apart from these benefits there is one other significant advantage – SDH networks are much easier to manage in operation. Partly this is due to the fact that SDH was conceived as a technology for a whole network (rather than a set of individual links), partly this is due to the fact that SDH is simply more modern, and therefore the available network management tools are more advanced.

SONET (Synchronous Optical Network)

SONET is the name of the North American variant of SDH. It is the forerunning technology which led to the ITU's development of SDH. The principles of SONET are very similar to those of SDH, but the terminology differs. The SONET equivalent of an SDH synchronous transfer module (STM) has one of two names, either *optical carrier (OC)* or *synchronous transport system (STS)*. The SONET equivalent of an SDH virtual container (VC) is called a *virtual tributary (VT)*. Some SDH STMs and VCs correspond

Table A1.1 Payload Rates of SDH Containers

Container type	Container frame size	Frame repetition rate	Capable of carrying PDH line type
C-11	193 bits	8000	T1 (1544 kbit/s)
C-12	256 bits	8000	E1 (2048 kbit/s)
C-21	789 bits	8000	T2 (6312 kbit/s)
C-22	1056 bits	8000	E2 (8448 kbit/s)
C-31	4296 bits	8000	E3 (34 368 kbit/s)
C-32	5592 bits	8000	T3 (44 736 kbit/s)
C-4	260 × 9 bytes	8000	139 264 kbit/s

exactly with SONET STS and VT equivalents. Some do not. The table of figure 11 presents a comparison of the two hierarchies.

Table A1.2 Comparison of SDH and SONET hierarchies

North American SONET	Carried Bitrate/Mbit/s	SDH
VT 1.5	1.544	VC-11
VT 2.0	2.048	VC-12
VT 3.0	3.152	–
VT 6.0	6.312	
VC-21	8.448	–
VC-22	34.368	–
VC-31	44.736	–
VC-32	149.76	VC-4
STS-1 (OC-1)	51.84	-
STS-3 (OC-3)	155.52	STM-1
STS-6 (OC-6)	311.04	-
STS-9 (OC-9)	466.56	-
STS-12 (OC-12)	622.08	STM-4
STS-18 (OC-18)	933.12	-
STS-24 (OC-24)	1244.16	-
STS-36 (OC-36)	1866.24	-
STS-48 (OC-48)	2488.32	STM-16
STS-96 (OC-96)	4976.64	-
STS-192 (OC-192)	9953.28	STM-64

Appendix 2
Overview of ATM Standards

This appendix provides a short glossary-style listing of the main ITU-T and ATM Forum specifications relating to ATM. The glossary is a listing of standards as published before December 1995. Unpublished standards still in production are, however, not listed.

The listing is intended to provide a simple guide to the standards, allowing interested readers to refer directly to the standards texts in order to obtain more detailed information not covered in this book. Figures A2.1 and A2.2 are additionally provided to remind the reader of the various interfaces defined by the ATM network reference model and of the relationship of the various signalling protocols to one another.

Figure A2.1 illustrates the interfaces specified by ATM Forum and the names assigned to the various interfaces. Not all are yet fully specified.

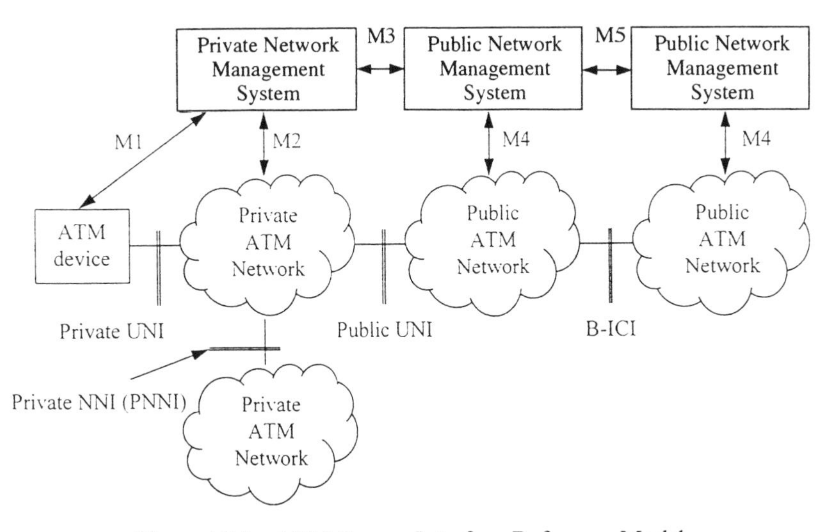

Figure A2.1 ATM Forum Interface Reference Model

Inter-Relationship of ATM Standards (Layers and Signalling Protocols)

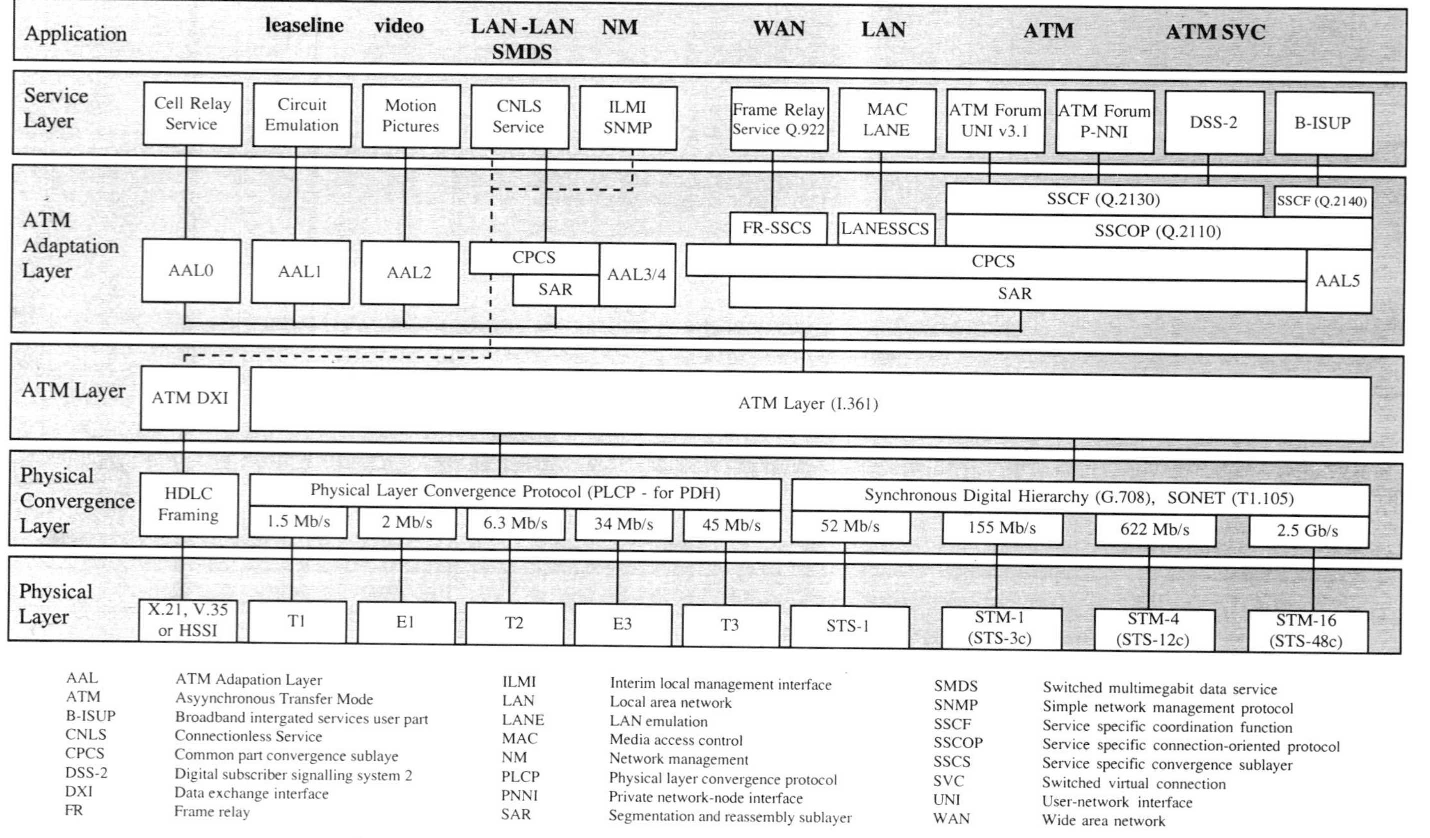

Figure A2.2 Interelationships of ATM layers and signalling standards

Standards Listing

Basic network principles of ATM (network aspects and functions)

Standard or recommendation (reference)	Title	Document status	Contents	Equivalent, similar or related standard
I.113 (ITU-T)	Vocabulary for B-ISDN	April 1991 original document 18 pages	Glossary of ATM terms and acronyms	
I.121 (ITU-T)	Broadband Aspects of ISDN	April 1991 original document 2 pages	Decribes the basic principles of B-ISDN	
I.150 (ITU-T)	B-ISDN Asynchronous Transfer Mode Functional Characteristics	July 1995 document rev.1 8 pages	Describes the basic technical features of the ATM layer, introducing Virtual Connections and Virtual Paths	ETSI prETS 300 298–1
I.211 (ITU-T)	General Service Aspects of B-ISDN	March 1993 original document 13 pages	A foundation document describing the basic services supported by B-ISDN (e.g. interactive service, distribution service, multimedia, video, etc.) and how they are impacted by ATM transmission	
I.311 (ITU-T)	B-ISDN General Network Aspects	March 1993 document rev.1 37 pages	Describes the basic network elements of a B-ISDN network (crossconnect, switch etc.), the interfaces (UNI, NNI) and how connections are established	
I.321 (ITU-T)	B-ISDN Protocol Reference Model and its Application	April 1991 original document 7 pages	Describes the Protocol Model and the functions of the various *layers* and *planes* within it	ETSI prETS 300 354

Standard or recommendation (reference)	Title	Document status	Contents	Equivalent, similar or related standard
I.327 (ITU-T)	B-ISDN Functional Architecture	March 1993 document rev.1 11 pages	Describes the basic architectural model for B-ISDN, defining network reference points and connection types and relating them to ISDN (i.e. connection-oriented) and connectionless services	
I.350 (ITU-T)	General Aspects of Quality of Service and Network Performance in Digital Networks, including ISDN	March 1993 document rev. 1 13 pages	Defines the general quality aspects of operating ISDN networks	
I.356 (ITU-T)	B-ISDN ATM Layer Cell Transfer Performance	November 1993 original document 19 pages	Defines the reference events which should be used to measure B-ISDN network performance	
I.361 (ITU-T)	B-ISDN ATM Layer Specification	July 1995 document rev.2 12 pages	Detailed specification of the 53-byte ATM cell structure and coding	ETSI prETS 300 298–2 ANSI T1.627
I.362 (ITU-T)	B-ISDN ATM Adaptation Layer (AAL) Functional Description	March 1993 document rev.1 3 pages	Short desription of the purpose of the ATM adaptation layer and definition of the various AAL service classes	
I.363 (ITU-T)	B-ISDN ATM Adaptation Layer (AAL) Specification	March 1993 document rev.1 68 pages; Addendum in November 1995 for AAL5	Detailed specification of the AAL and its coding	ETSI ETS 300 353 AAL1 ETS 300 349 (AAL3/4) ETS 300 428 (AAL5) ANSI T1.629, T1.630, T1.635

I.364 (ITU-T)	Support of Broadband Connectionless Data Service on B-ISDN	July 1995 document rev.1 10 pages	Describes the framework for support of BCDBS (broadband connectionless data bearer service) and the protocols to be used at UNI and NNI	
I.365.1 (ITU-T)	Frame Relaying Bearer Service Specific Convergence Sublayer (FR-SSCS)	November 1993 original document 8 pages	Detailed specification for the interworking of frame relay with ATM	ANSI T1.634
I.365.2 (ITU-T)	Service specific coordination function (SSCF) to provide the connection-oriented network service (CONS)	July 1995 original document	Describes the mapping of control primitives from the synchronization and coordination function (SCF) of ITU-T Recommendation Q.923 to the signals of SSCOP for support of CONS	
I.365.3 (ITU-T)	Service specific coordination function (SSCF) to provide the connection-mode oriented transport service (COTS)	July 1995 original document	Describes the mapping of control primitives from the synchronization and coordination function (SCF) of ITU-T Recommendation Q.923 to the signals of SSCOP for support of COTS	
I.371 (ITU-T)	Traffic Control and Congestion Control in B-ISDN	March 1993 original document 26 pages	Description of the methodology for controlling traffic quality by cell discarding and rejection of new connections	ETSI prETS 300 301
I.374 (ITU-T)	Framework Recommendations on network capabilities to support multimedia service	March 1993 original document 7 pages	A foundation document introducing the concept of multimedia services and the network attributes (i.e. B-ISDN) needed to support them	

ATM User-Network Interface (UNI)

Standard or recommendation (reference)	Title	Document status	Contents	Equivalent, similar or related standard
I.411 (ITU-T)	ISDN User-Network Interfaces – Reference configurations	March 1993 document rev.1 7 pages	Defines the various terminology and reference points of the UNI (R, S, T, B-NT1, B-NT2, B-TE, etc.)	
I.413 (ITU-T)	B-ISDN User Network Interface (UNI)	March 1993 document rev.1 9 pages	Defines the terminology for end equipment (NT, TE) connected to ATM networks and the various connection reference points (R, S, T, W)	ATM Forum UNI ETSI prETS 300 299 (cell based UNI) pr ETS 300 300 (SDH based UNI)
I.414 (ITU-T)	Overview of Recommendations on Layer 1 for ISDN- and B-ISDN Customer Accesses	March 1993 original document 5 pages	Overview document providing reference to relevant G and Q series recommendations	ETSI prETS 300 299 (cell based UNI) prETS 300 300 (SDH based UNI)
I.432 (ITU-T)	B-ISDN User Network Interface (UNI) – Physical Layer Specification	March 1993 document rev.1 36 pages	Detailed definition of electrical and optical interfaces to be used at the UNI	ETSI prETS 300 299 (cell based UNI) prETS 300 300 (SDH based UNI)

Interworking with other networks and services

Standard or recommendation (reference)	Title	Document status	Contents	Equivalent, similar or related standard
I.501 (ITU-T)	ISDN Internetwork interfaces – service interworking	March 1993 original document 8 pages	Provides some information on bearer service interworking, including frame mode bearer	
I.555 (ITU-T)	Frame Relay Bearer Service Interworking	November 1993 original document 21 pages	Defines general interworking of Frame relay with other networks, including B-ISDN	ANSI T1.633
I.580 (ITU-T)	General Arrangements for Interworking between B-ISDN and 64 kbit/s based ISDN	July 1995 document rev.1 21 pages	Defines general interworking between *narrowband* ISDN and B-ISDN	

Operations and maintainance of B-ISDN

Standard or recommendation (reference)	Title	Document status	Contents	Equivalent, similar or related standard
I.610 (ITU-T)	B-ISDN Operation and Maintenance (OAM) Principles and Functions	July 1995 document rev.1 46 pages	Describes operational procedures for maintaining B-ISDN networks93	

Broadband bearer services

Standard or recommendation (reference)	Title	Document status	Contents	Equivalent, similar or related standard
F.811 (ITU-T)	Broadband connection oriented bearer (BCOB) service	August 1992 original document 10 pages	Defines the broadband connection oriented bearer service (BCOB) and its various classes	
F.812 (ITU-T)	Broadband connectionless data bearer service (BCDBS)	August 1992 original document 6 pages	Defines the broadband connectionless data bearer service (BCDBS)	
F.813 (ITU-T)	Virtual path service for reserved and permanent communications	February 1995 original document	Defines the broadband virtual path service	

Signalling, setting up switched connections and supplementary services

Standard or recommendation (reference)	Title	Document status	Contents	Equivalent, similar or related standard
Q.704, Q.707 (ITU-T)	Message Transfer Part Level 3 (MTP3)	March 1993 document rev.1	Defines the message transfer part protocol (MTP3) which underlies both N-ISUP and B-ISUP	
Q.782 (ITU-T)	MTP Level 3 Test Specification	March 1993 document rev.1	Defines a testing procedure for validating MTP3	
Q.940 (ITU-T)	UNI Protocol for ISDN management – general aspects	*Blue Book*, 1988	Defines the basic principles of signalling layer management	
Q.2010 (ITU-T)	B-ISDN overview – signalling capability set 1, release 1	February 1995 original document	Describes signalling reference configurations, relationships and protocol stacks to be used in B-ISDN	
Q.2100 (ITU-T)	B-ISDN signalling ATM adaptation layer (SAAL) overview description	July 1994 original document 3 pages	Describes briefly the AAL functions necessary to support signalling (SAAL)	
Q.2110 (ITU-T)	B-ISDN ATM adaptation layer – service specific connection oriented protocol (SSCOP)	July 1994 original document 95 pages	Describes the SSCOP sublayer of the SSCS (service specific convergence sublayer). SSCS is itself a sublayer of SAAL. Used in conjunction with either Q.2130 or Q.2140	ETS 300 436-1
Q.2120 (ITU-T)	B-ISDN meta-signalling protocol	February 1995 original document	Defines the meta-signalling (bootstrap) activity which allows B-ISDN terminals to request two signalling VCs at the UNI	

Standard or recommendation (reference)	Title	Document status	Contents	Equivalent, similar or related standard
Q.2130 (ITU-T)	B-ISDN signalling ATM adaptation layer – service specific coordination function (SSCF) for support of signalling at the UNI	July 1994 original document 54 pages	Defines the SSCF at the UNI. Used in conjunction with SSCOP (Q.2110) to form the UNI SSCS (service specific convergence sublayer)	ETS 300 437-1 ETS 300 437-2
Q.2140 (ITU-T)	B-ISDN ATM adaptation layer – service specific coordination function (SSCF) at the NNI	February 1995 original document	Defines the SSCF at the NNI. Used in conjunction with SSCOP (Q.2110) to form the UNI SSCS (service specific convergence sublayer)	ETS 300 438-1
Q.2144 (ITU-T)	B-ISDN signalling ATM adaptation layer – layer management for the SAAL at the NNI	May 1995 original document	Defines the management coordination and error monitoring of SAAL at the NNI	
Q.2610 (ITU-T)	B-ISDN usage of cause and location in B-ISUP and DSS2	February 1995 original document	Defines the mapping tables and interworking for cause and location signals used during basic SVC call setup and cleardown in B-ISUP and DSS2	
Q.2650 (ITU-T)	Interworking between B-ISUP and DSS2	February 1995 original document	Defines the mapping tables and interworking of signals used during basic SVC call setup and cleardown in B-ISUP and DSS2	
Q.2660 (ITU-T)	Interworking between B-ISUP and N-ISUP	February 1995 original document	Defines the mapping tables and interworking of signals used during basic SVC call setup and cleardown in B-ISUP and N-ISUP (in a narrowband ISDN)	

Q.2730 (ITU-T)	B-ISUP supplementary services	February 1995 original document	Defines the supplementary services of B-ISUP capability set 1 (Q.2010)
Q.2761 (ITU-T)	Functional description of B-ISUP of SS7	February 1995 original document	Describes the basic funtionality, structure and capabilities of B-ISUP
Q.2762 (ITU-T)	General functions of messages and signals of the B-ISUP of SS7	February 1995 original document	Describes the basic signalling information in B-ISUP
Q.2763 (ITU-T)	B-ISUP – formats and codes	February 1995 original document	Defines the detailed coding of signals for B-ISUP
Q.2764 (ITU-T)	B-ISUP – basic call procedures	February 1995 original document	Defines the detailed signalling procedures between network nodes (i.e. at the NNI) for basic call setup and cleardown of SVCs in B-ISDNs
Q.2931 (ITU-T)	Digital subscriber signalling number 2 (DSS2) – UNI layer 3 specification for basic call/connection control	February 1995 original document	Defines the signalling and procedures of DSS2 used at the UNI for establishing SVCs
Q.2951 (ITU-T)	Stage 3 description for number identification supplementary services using DSS2 – basic call	Clause 1 – direct dialling in (February 1995) Clause 2 – multiple subscriber number (February 1995) Clause 3 – call identification (February 1995) Clause 4 – call non-identification (February 1995) Clause 5 – connected line identification (February 1995) Clause 6 – connected line non-identification (February 1995) Clause 8 – sub-addressing (February 1995)	Defines the operation of DSS2 for support of B-ISDN supplementary services

Standard or recommendation (reference)	Title	Document status	Contents	Equivalent, similar or related standard
Q.2957 (ITU-T)	Stage 3 description for additional information transfer supplementary services using DSS2 – basic call	Clause 1 – user-to-user signalling (UUS) (February 1995)	Defines the operation of DSS2 for support of user-to-user signalling (UUS)	
Q.2961 (ITU-T)	DSS2 – support of additional parameters	May 1995 original document	Defines the use and carriage of parameters by DSS2 at the UNI to enable control of traffic and bandwidth allocation at T_B and S_B interfaces	
Q.2971 (ITU-T)	DSS2 – UNI layer 3 specification for point-to-multipoint call/connection control	May 1995 original document	Defines the setting-up and use of SVCs in B-ISDNs for multipoint connections, including how to add and remove parties during the call	

Transmission media suitable for ATM

Standard or recommendation (reference)	Title	Document status	Contents	Equivalent, similar or related standard
G.652 (ITU-T)	Characteristics of a single mode (monomode) optical fibre cable	March 1993 document rev.1 8 pages	Description of monomode fibre, as used in telecom networks, including ATM	
G.708 (ITU-T)	Network node interface for the Synchronous Digital Hierarchy	March 1993 document rev.2 16 pages	Describes SDH technologies, including how to map ATM cells into the SDH containers	
G.782 (ITU-T)	Types and General Characteristics of SDH multiplexing equipment	January 1994 document rev.1 29 pages	Overview of the functions of SDH transmission equipment, multiplexors and crossconnects	
G.804 (ITU-T)	ATM Cell Mapping into Plesiochronous Digital Hierarchy (PDH)	November 1993 original document 19 pages	Describes how ATM cells may be mapped into PDH transmission	
G.805 (ITU-T)	Generic functional architecture of transport networks	July 1995 original document	Lays out the future network architecture for transport networks in general	
G.806 (ITU-T)	Functional architecture of transport networks based on ATM	July 1995 original document	Based upon G.805, this recommendation defines the architecture of transport networks based on ATM	
G.902 (ITU-T)	Framework recommendation on Functional access networks	July 1995 original document	Describes the network functions and architecture of future networks with particular regard to the network access requirements. The network access architecture and 'service node interfaces' for future access types including broadband services This is the basis of a common interface between ATM and physical layers in ATM equipment	

ATM Forum specifications

Standard or recommendation (reference)	Title	Document status	Contents	Equivalent, similar or related standard
Network interfaces				
UNI	User–Network Interface	v3.0 (1993) v3.1 (1995)	Specifies all aspects of the UNI (physical and protocol layers)	
P-NNI	Private–Network-Node Interface	draft	Specifies the P-NNI protocol for use between private ATM switches. The protocol encompasses two elements: one for distributing information about the network topology, the second for passing signalling information for setting up connections	
IISP	Interim Inter-Switch Signalling Protocol (IISP) specification	v1.0 (December 1994)	Specifies the interim inter-switch protocol to be used at the PNNI between private ATM networks	
B-ICI	Broadband-ISDN (B-ISDN) Inter-Carrier Interface (B-ICI) specification	v1.1 (September 1994)	Specifies the inter-carrier interface for use between public ATM networks. The ICI is an extension of the basic NNI. The higher layer protocols have been extended to cater for the special demands of inter-carrier demarcation	
Physical Layer Interfaces				
155 Mb/s Cat.5 TP	ATM physical medium dependent interface specification for 155 Mb/s	v1.0 (September 1994)	An extension to UNI v3.0 enabling the use of TP (twisted pair) cable as the physical medium for speeds up	

52 Mb/s Cat. 3 UTP	Mid-range physical layer specification for category 3 unshielded twisted pair (UTP)	v1.0 (September 1994)	An extension to UNI v3.0 enabling the use of UTP (unshielded twisted pair) cable as the physical medium for speeds up to 52 Mbit/s
UNI (6312 kbps)	ATM physical medium dependent interface specification for DS2 physical layer interface	v1.0	Specifies the physical layer interface for ATM at 6.312 Mbit/s
DS1 PHY	ATM physical medium dependent interface specification for DS1 physical layer interface	v1.0	Specifies the physical layer interface for ATM at 1.544 Mbit/s
ATM: network applications			
DXI	ATM Data-Exchange -Interface (DXI) specification	v1.0 (August 1993)	Specifies the ATM DXI. This is an interface allowing a pre-ATM equipment such as a router to cooperate with an ATM line unit or DCE to provide a UNI for ATM networks
LAN emulation	LAN emulation over ATM	v1.0 (January 1995)	Defines and specifies the use of ATM as an alternative to existing Token Ring and Ethernet LAN technology
CNM for ATM PNS (M3 spec)	Customer Network Management (CNM) for ATM public network service (M3 specification)	October 1994	Defines a customer network management (CNM) service to be offered by public ATM network providers
M4-interface and MIB	M4 Interface requirements and logical MIB	October 1994	Specifies the functional interfaces required to manage ATM networks
UTOPIA level 1	Utopia Specification	v2.01 (March 1994)	Specifies the *Universal Test and Operations Physical Interface for ATM (UTOPIA).*

Standard or recommendation (reference)	Title	Document status	Contents	Equivalent, similar or related standard
Testing and conformance				
Test specs introduction	Introduction to ATM Forum test specifications	v1.0 (December 1994)	An overview of the various testing areas and test specification types, e.g. PICS (protocol implementation conformance statement) and PIXITs (protocol implementation extra information for testing)	
PICS for STS-3c PHY	PICS Proforma for the SONET STS-3c physical layer interface	v1.0	PICS test specification for SONET STS-3c (155 Mb/s)	
PICS for 100 Mb/s fibre	PICS Proforma for the 100 Mbps Multimode fibre physical layer interface	v1.0	PICS test specification for multimode fibre interface (100 Mb/s – as used by FDDI networks previously)	
PICS for DS3 PHY	PICS Proforma for the DS3 physical layer interface	v1.0	PICS test specification for DS3 interface (45 Mb/s)	

Glossary of Terms

A	Applicable
AA	Administrative authority (for addressing)
AAL	ATM Adaptation Layer
AAL-IE	AAL information element
AAL-PCI	AAL protocol control information
AAL-S D U	AAL service data unit
AAL1	AAL type 1 (constant bit rate service)
AAL2	AAL type 2 (variable bit rate, but time critical service)
AAL3/4	AAL type 3/4 (class C frame relaying service or class D connectionless service)
AAL5	AAL type 5 (class C frame relaying service or class D connectionless service)
ABM	]Asynchronous balanced mode (HDLC)
ABR	Available bit rate
ACE	Access connection element
ACF	Access control field (DQDB)
ACK	Acknowledgement
ACR	Available cell rate
ADM	Add/drop multiplexor (SDH)
AFI	Address format identifier or authority and format identifier
AII	Active input interface (optical fibre)
AIM	ATM inverse multiplexing
AIS	Alarm indication signal
AL	Alignment (AAL3/4 CPCS)
AL	Access link
AM	Accounting management
AMS	Audio visual multimedia services
AN	Access Node
ANSI	American National Standards Institute
AOI	Active output interface (optical fibre)
AP	Acknowledgement packet
API	Application programming interface
APS	Automatic protection switching
ARE	All routes explorer
ARP	Address resolution protocol

ATE	ATM terminating equipment
ATM	Asynchronous transfer mode
ATM traffic descriptor	The generic list of traffic parameters that define the characteristics of an ATM connection
ATM-SDU	ATM service data unit
ATMM	ATM management
AU	Administrative unit (SDH terminology)
AUI	Attachment unit interface (ethernet LAN)
AUU	ATM layer user-to-user indication
AVI	Audiovisual interactive service
AW	Administrative weight
B-BC	Broadband bearer capability
B-HLI	Broadband higher layer information
B-ICI	Broadband inter carrier interface
B-ISDN	Broadband integrated services digital network
B-ISDN PRM	Protocol reference model for B-ISDN
B-ISPBX	Private branch exchange for B-ISDN
B-LLI	Broadband lower layer information
B-NT	Network Termination for B-ISDN
B-NT1	Network Termination 1 for B-ISDN
B-NT2	Network Termination 2 for B-ISDN (multipoint configuration)
B-SP	B-ISDN signalling point
B-STP	B-ISDN signalling transfer point
B-TA	Terminal Adaptor for B-ISDN
B-TE	Terminal Equipment for B-ISDN
B-TE1	B-ISDN Terminal Equipment type 1 (designed for ATM)
B-TE2	B-ISDN Terminal equipment type 2 (connected via B-TA)
BASize	Buffer allocation size (AAL 3/4 CPCS)
Bc	Committed burst (frame relay)
BCD	Binary coded decimal
BCDBS	Broadband connectionless data bearer service
BCOB	Broadband connection-oriented bearer (service)
BCOB-A	Broadband connection-oriented bearer – A: constant bit rate ATM transport with stringent end-to-end timing needs
BCOB-C	Broadband connection-oriented bearer – C: variable bit rate ATM transport with no end-to-end timing needs
BCOB-X	Broadband connection-oriented bearer- X: ATM-only: network will not process layers higher than ATM layer
Be	Excess burst (frame relay)
BECN	Backward explicit congestion notification (frame relay)
BEDC	Block error detection code (ATM OAM performance cell)
BER	Bit error ratio (or bit error rate)
BGP	Border gateway protocol
BIP	Bit interleaved parity code (block error detection code)
BIP	Bit interleaved parity
BIS	Border intermediate system
BIS PDU	Border intermediate system protocol data unit
BISSI	Broadband inter-switching system interface
BLER	Block error result (ATM OAM performance cell)

block	A unit of information consisting of a *header* and an information field
block payload	The bits in the information field of a block
BN	Bridge number
BOM	Beginning of message
BPP	Bridge port pair (service routing descriptor)
BRI	Basic rate interface (ISDN)
broadband	A service or system supporting rates greater than 2 Mbit/s
broadcast	A service providing unidirectional distribution to multiple receivers
Btag	Beginning tag (AAL3/4 CPCS)
BUS	Broadcast and unknown server
BVPS	Broadband virtual path services
BVPS-P	BVPS for permanent communications
BVPS-R	BVPS for reserved communications
C (with suffix VCI or VPI)	Control function for a virtual connection of virtual path
C-n	Container-n (SDH terminology)
C/R	Command or Response bit
CA	Customer access (TMN)
CAC	Connection admission control
CAD-CAM	Computer aided design/Computer aided manufacturing
CAMC	Customer access maintenance centre
CAMF	Customer access management function (TMN)
CAP	Carrierless amplitude modulation/phase modulation
CBDS	Connectionless broadband data service
CBR	Constant bit rate
CCITT	International telephone and telegraph consultative committee (now ITU-T – the telecommunication standardization sector of ITU)
CD	Countdown counter (DQDB)
CDV	Cell delay variation
CDVT	Cell delay variation tolerance
CE	Connection endpoint
CE	Connection element
CED	Cell error control (physical layer)
CEI	Connection endpoint identifier
cell	A *block* of fixed length (48 byte information field and 5 byte header)
cell identification	The identification of cell boundaries in a cell or bit stream
cell tagging	A process in which the cell loss priority (CLP) is changed by the network from 'O' to '1', allowing congestion relief by cell discarding
CEQ	Customer Equipment
CER	Cell error ratio
CES	Connection endpoint suffix
CES	Circuit emulation service
CI	Customer installation (TMN)
CI	Congestion indication
CIB	CRC-32 indicator bit
CIME	Customer installation maintenance entities

CIMF	Customer installation management function (TMN)
CIR	Committed information rate (frame relay)
circuit transfer mode	A telecommunication transfer requiring permanent allocation of bandwidth
CL	Connectionless
Clav	Cell available (flow control)
Clk	Clock
CLLM	Consolidated link layer management (FR)
CLNAP	Connectionless network service access protocol
CLNS	Connectionless network service (OSI)
CLP	Cell loss priority
CLR	Cell loss ratio
CLSF	Connectionless service function
CM	Configuration management
CME	Connection management entity
CMI	Coded mark inversion (line code at the physical layer when coaxial cable is used)
CMIP	Common management information protocol
CMIS	Common management information service
CMISE	Common management information service element
CMR	Cell misinsertion rate or ratio
CN	Customer network
CN	Congestion notification
CNM	Customer network management
CNR	Complex node representation
CO	Connection oriented
COH	Connection overhead
COM	Continuation of message
CON	Concentrator
connection admission control	The procedure by which a network decides whether to accept a new connection
Connection traffic descriptor	Specifies the traffic characteristics of an ATM connection at the public or private UNI
connectionless service	A service allowing for information transfer without end-to-end call establishment
CONS	Connection-oriented network service (OSI)
constant bit rate	A service characterized by a service bit rate of constant value
contribution	A broadband service used to submit a vidio signal for post production processing
conversational servce	An interacfive service providing for bidirectional communication
COTS	Connection-oriented transport service
CPCS	Common part convergence sublayer (AAL3/4 and AALS)
CPCS-CI	CPCS-congestion indication
CPCS-LP	CPCS-loss priority
CPC-UU	CPCS-user-to-user indication
CPE	Customer premises equipment
CPI	Common part indicator (CPCS AAL3/4 or AALS)
CPN	Customer premises network
CR	Conditional requirement

CRC	Cyclic redundancy check
CRF	Connection related function
CRF (VC)	Virtual channel connection related function
CRF (VP)	Virtual path connection related function
CRM	Cell rate margin
CRS	Cell relay service
CS	Convergence sublayer
CS-PDU	Convergence sublayer protocol data unit
CSI	Convergence sublayer indication bit (AAL1)
CSR	Cell missequenced ratio
Ctag	Confirmation tag
CTD	Cell transfer delay
CV	Code violation
D	Direction
D/C	DLCI or DL-CORE control indicator (frame relay or FR-SSCS)
DA	Destination address field
DAL	Dedicated access line
DCC	Data country code
DCE	Data circuit terminating equipment
DE	Discard eligibility (frame relay or FR-SSCS)
DFA	DXI frame address
DFI	DSP format identifier (address)
DH	DMPDU header (DQDB)
distribution	A broadband service used to distribute video or audio signals to end users
DLCI	Data link connection identifier (frame relay)
DMPDU	Derived MAC (media access control) PDU (DQDB)
DPL	Primary link for distribution services
DQDB	Distributed queue dual bus (the technology behind SMDS)
DS	Digital Section (transmission line)
DSO	Digital section level O (North American standard – 64 kbit/s)
DS1	Digital section level 1 (North American standard – 11 – 1544 kbit/s)
DS3	Digital section level 3 (North American standard -13 – 45 Mbit/s)
DSG	Default slot generator (DQDB)
DSP	Domain specific part (address)
DSS	Distributed samble scrambler
DSS1	Digital signaling system system 1 (narrow bond ISDN)
DSS2	Digital signalling system 2 (broadband ISDN)
DSU	Digital service unit (digital line terminating device)
DT	DMPDU trailer (DQDB)
DTE	Data terminal equipment
DTL	Designated transit list
DTP	Data transfer protocol
DXC	Digital crossconnect
DXI	Data exchange interface
E1	Digital section level 1 (European standard – 2048 kbit/s)
E3	Digital section level 3 (European standard – 34 Mbit/s)

EA	Extended address
EA	Address extension bit
EBCN	Explicit backward congestion notification
EC	Error correction
ED	Error detection
EDC	Error Detection Code (physical layer)
EFCI	Explicit forward congestion indication
EFCN	Explicit forward congestion notification
EFD	Event forwarding discriminator
EFI	Errored frame indicator
EGP	Exterior gateway protocol
EGRP	Exterior gateway routing protocol
EIA	Electrical industries association
EIR	Excess information rate (FR)
EL	Element layer (TMN)
EL	Electrical (interface)
ELAN	Emulated local area network
EMC	Electromagnetic compatibility
EML	Element management layer (TMN)
Enb	Enable
EOM	End of message
EOT	End of transmission
ERA	Exterior reachable ATM address
ES	End system
ES	Errored seconds
ESI	End system identifier
ESID	End system identifier
ESIG	European SMDS interest group
ET	Exchange termination
Etag	End tag (AAL3/4 CPCS)
ETSI	European Telecommunications Standards Institute
F0 state	Loss of power on the user side of the UNI (physical layer)
F1 cell	Physical layer OAM cell (regenerator level)
F1 flow	Protocol communication at the regenerator section level (ATM physical layer)
F1 level	Regenerator section level (ATM physical layer)
F1 state	Operational state on the user side of the UNI (physical layer)
F2 cell	Physical layer OAM cell (digital section level)
F2 flow	Protocol communication at the digital section level (ATM physical layer)
F2 level	Digital section level (ATM physical layer)
F2 state	Fault condition number 1 (FC1) on the user side of the UNI (physical layer)
F3 cell	Physical layer OAM cell (transmission path level)
F3 flow	Protocol communication at the transmission path level (ATM physical layer)
F3 level	Transmission path level (ATM physical layer)
F3 state	Fault condition number 2 (FC2) on the user side of the UNI (physical layer)
F4 flow	Protocol communication at the virtual path level (ATM layer)

F4 level	Virtual path level (ATM layer)
F4 state	Fault condition number 3 (FC3) or FC1 and FC3 on the user side of the UNI (physical layer)
F5 flow	Protocol communication at the virtual channel level (ATM layer)
F5 level	Virtual channel level (ATM layer)
F5 state	Fault condition number 4 (FC4) on the user side of the UNI (physical layer)
F6 state	FC3=FC4 or FC1=FC3=FC4 on the user side of the UNI (physical layer)
F7 state	Power on state (transient state) at the user side of the UNI (physical layer)
FCn	Fault condition number n (physical layer) – the various fault conditions are defined in ITU-T 1.432
FCS	Frame check sequence (FR)
FDDI	Fibre distributed data interface
FEA	Functional entity actions
FEAC	Far end alarm and control
FEBE	Far end block error
FEC	Forward error correction
FECN	Forward explicit congestion notification (frame relay)
FERF	Far end receive failure (subsequently renamed RDI, remote defect indication)
FIFO	First in first out (queueing mechanism)
FM	Fault management
FMBS	Frame mode bearer service
FR	Frame relay (correctly *frame relaying*)
FR-IWP	Frame relaying interworking point
FR-SSCS	Frame relaying service specific convergence sublayer
frame	A *block* of variable length identified by a header label
FRBS	Frame relaying bearer service
FRM	Fast resource management
FRS	Frame relay service
FSBS	Frame switching bearer service
Full	Full availability
FUNI	ATM Frame user network interface
G0 state	Loss of power on the user side of the UNI (physical layer)
G1 state	Operational state on the network side of the UNI (physical layer)
G10 state	F1=FC4 or FC2=FC3=FC4 on the network side of the UNI (physical layer)
G11 state	FC1=FC2=FC3 on the network side of the UNI (physical layer)
G12 state	FC1=FC3=FC4 or FC1=FC2=FC3=FC4 on the network side of the user interface (physical layer)
G13 state	Power-on state transient state) on the network side of the UNI (physical layer)
G2 state	Fault condition number 1 (FCl) on the network side of the UNI (physical layer)
G3 state	Fault condition number 2 (FC2) on the network side of the UNI (physical layer)

G4 state	Fault condition number 3 (FC3) on the network side of the UNI (physical layer)
G5 state	FC4 or FC4=FC2 on the network side of the UNI (physical layer)
G6 state	FC1=FC2 conditions on the network side of the UNI (physical layer)
G7 state	FC1=FC3 conditions on the network side of the UNI (physical layer)
G8 state	FC1=FC4 or FC1=FC2=FC4 conditions on the network side of the UNI (physical layer)
G9 state	FC2=FC3 conditions on the network side of the UNI (physical layer)
GBSVC	General broadcast signalling virtual channel
GCAC	Generic connection admission control
GCRA	Generic cell rate algorithm
GFC	Generic flow control
GME	Global management entity
GPN	Generating polynomial (network)
GPU	Generating polynomial (user)
H	Header
HCS	Header check sequence (DQDB)
HDLC	High level datalink control (OSI)
Hdr	Header
HDTV	High definition television
header	The bits within a block or cell which provide for correct delivery of the payload
HEC	Header error control
HLF	Higher layer function
HLI	Higher layer information
HOB	Head of bus (DQDB)
HSSI	High speed serial interface (EIA/TIA 612/613)
Ia	Interface point on the user terminal side of the SB-reference point at the UNI
IA5	International alphabet number 5 (the ASCII code)
Ib	Interface point on the B-NT2 side of the SB-reference point at the UNI
ICD	International code designation (address)
ICI	Inter-carrier interface
ICIP	Inter-carrier interface protocol
ICIP—CLS	ICIP connectionless services
ICMP	Internet control message protocol
ID	Interface data (AAL3/4 SAR-SDU)
IDI	Initial domain identifier (address)
IDP	Initial domain part
IDU	Interface data unit
IE	Information element
IEC	Inter-exchange carrier (USA)
IEEE	Institute of Electrical and Electronics Engineers
IETF	Internet engineering task force
IGP	Interior gateway protocol
IISP	Interim interswitch signalling protocol (ATM NNI)

ILEC	Independent local exchange carrier (USA)
ILMI	Interim local management interface (based on SNMP)
IMPDU	Initial MAC (media access control) PDU (DQDB)
IMSSI	Inter-MAN (metropolitan area network) switching system interface (DQDB)
IND	Indicate
Info	CPCS payload
INI	Inter-network interface
interactive service	The means used to interchange data between systems (either a storage or transmission means or both)
invalid cell	A cell where the header is declared by the header error control process to contain errors
IOP	Interoperability testing
IP	Internet protocol
IPI	Initial protocol identifier (ISO/IEC TR 9577)
IPL	Primary link for interactive services
IRA	Internal reachable ATM address
IRP	internal reference point
IS	Intermediate system
ISDN	Integrated services digital network
ISDU	Isochronous service data unit (DQDB)
ISO	International Organization for Standardization
ISP	Interswitch signalling protocol (ATM NNI)
ISSI	Inter-switching system interface
ISUP	ISDN-user part (signalling system number 7)
IT	Information type
ITU	International Telecommunications Union
ITU-T	The telecommunications standardization sector of ITU (former CCITT)
IUT	Implementation under test
IWF	Interworking function
IWP	Inter-working point
kbit/s	kilobits per second
Kbps	kilobit per second
Kx	Inter-network reference point – the point between a B-ISDN and some other sort of transit network
labelled channel	A temporary collection of all block payloads having a common label value
Lan	Local Area Network
LANE	LAN emulation
LAP	Link access procedure (OSI)
LAPB	Balanced link access procedure (X.25)
LAPD	ISDN link access procedure (D-channel)
LC-node	Last common node
LCGN	Logical channel group number
LCT	Last compliance time or last conformance time (leaky bucket algorithm)
LD	LAN destination
LE	Local exchange
LE	Local exchange
LE	Layer entity

LE	LAN emulation
LE—ARP	LAN emulation address resolution protocol
LEC	Local exchange carrier (USA)
LEC	LAN emulation client
LECID	LAN emulation client identifier
LECS	LAN emulation configuration server
LF	Largest framesize
LFC	Local function capabilities
LGN	Logical group node
LI	Link identifier
LI	Length indicator
LLC	Lower layer compatibility (ISDN)
LLC	Logical link control (token ring and ethernet LANs)
LLI	Lower layer information
LME	Layer management entity
LMI	Local management interface
LOC	Loss of cell delineation
LOF	Loss of frame
logical signalling channel	A logical channel used to carry signalling information
LOM	Loss of OAM cell
LOP	Loss of pointer
LSAP	Link service access point
LSB	Least significant bit
LT	Line termination
LT	Line termination
LTE	Line terminating equipment
LTH	Length field
LUNI	LAN-user-network-interface
M	More (AAL3/4 SAR-SDU)
M reference point	Interface point between a B-ISDN and a CLSF (control function) of an external specialized service provider
M1-interface	Interface between ATM end device and private ATM network management system (ATM forum)
M2-interface	Interface between Private ATM network and its management system (ATM forum)
M3-interface	Interface offered from public ATM network management system to enable customer network management (ATM forum)
M4-interface	Interface between Public ATM network and its management system (ATM forum)
M5-interface	Interface between management systems of different Public ATM networks (ATM forum)
MA	Medium Adaptor
MAC	Media access control (ethernet and token ring LANs)
MAN	Metropolitan area network
MaxCR	Maximum cell rate
Mb/s	Megabits per second
Mbit/s	Megabits per second
Mbps	Megabits per second
MBS	Monitoring block size (PL-OAM cell)

MBS	Maximum burst size
MCD	Maintenance cell description
MCDV	Maximum cell delay variation
MCF	MAC convergence function (DQDB)
MCLR	Maximum cell loss ratio
MCP	MAC convergence protocol (DQDB)
MCR	Maximum cell rate
MCSN	Monitoring cell sequence number (ATM OAM performance management cell)
MCTD	Maximum cell transfer delay
media	Plural of *medium*
medium	A means by which information can be perceived, expressed, stored or transmitted
messaging service	An *interactive* service based upon store-and-forward and mailbox transfer
meta-signalling	The procedure for establishing, checking and releasing signalling virtual channels
MF	Management function (TMN)
MF	Mapping function
MIB	Management information base (ILMI/SNMP)
MIC	Media interface connector (jack, plug or socket)
MID	Multiplexing identification (ATM)
MID	Message identifier (DQDB)
mixed document	A document containing a mixture of text, graphics, data, image and moving pictures
ML	Maximum length
MM	Message mode (AAL3/4 or AALS)
MM	Multimode fibre
MP	Measurement point
MPEG	Motion Pictures Experts Group
MPH	Management physical header(physical layer primitive)
MPH-AI	MPH activate indication
MPH-CIn	MPH correction indication with parameter n
MPH-DI	MPH deactivate indication
MPH-EIn	MPH error indication with parameter n
MPOA	Multiprotocol over ATM
MS	Multiplex section (SDH)
MSAP	MAC Service access point (SMDS)
MSB	Most significant bit
MSDU	MAC service data unit (SMDS)
MSP	Management service provider (TMN)
MSP	Maintenance service provider
MSS	MAN switching system (SMDS)
MSVC	Meta signalling virtual channel
MTP	Message transfer part
MTP3	Message transfer part layer 3 (signalling system number 7)
MTU	Message transfer unit
multimedia service	A service in which interchanged information is a mixture of text, sound, graphics, video
multipoint	A communication or network configuration involving more than two end points

MUX	Multiplexor
N-ISDN	Narrowband ISDN
N/A	Not applicable
N0	(Call state at the user side of the UNI interface) null – no call exists
N1	(Call state at the network side of the UNI interface) call initiated
N10	(Call state at the network side of the UNI interface) active
N11	(Call state at the network side of the UNI interface) release request
N12	(Call state at the network side of the UNI interface) release indication
N22	(Call state at the network side of the UNI interface) call abort
N3	(Call state at the network side of the UNI interface) outgoing call proceeding
N4	(Call state at the network side of the UNI interface) call delivered
N6	(Call state at the network side of the UNI interface) call present
N7	(Call state at the network side of the UNI interface) call received
N8	(Call state at the network side of the UNI interface) connect request
N9	(Call state at the network side of the UNI interface) incoming call proceeding
NA	Network Adaptor (an interworking function between B-ISDN and narrowband ISDN)
NDF	New data flag
NDIS	Network driver interface specification
NE	Network element
Network Node Interface	The interface at a network node which is used to connect to another network node
NEXT	Near end crosstalk
NIC	Number of included cells (PL-OAM cell)
NIR	NEXT (near end crosstalk) loss-to-insertion ratio
NIU	Network interface unit
NLPID	Network layer protocol identifier (ISO/IEC TR 9577)
NM	Network management
NMB-EB	Number of monitored blocks – far end block errors (physical layer)
NMB-EDC	Number of monitored blocks – for which error detection codes are included (physical layer)
NML	Network management layer (TMN)
NNI	Network node interface
non-P-format	A convergence sublayer format type used by AAL1
NP	Network performance
NPC	Network parameter control
NPDU	Network protocol data unit
NRM	Network resource management
NTR	Not real time

NRZ	Non return to zero code (line code at the physical layer when optical fibre is used)
NSAP	Network service access point
NSR	Non-source routed
NT	Network termination
O	Optional
OAM	Operation and maintenance
OAMC	Operation and maintenance centre
OAMC	Operation administration maintenance centre (TMN)
OC-n	Optical carrier level-n (SONET)
OCD	Out of cell delineation (an anomaly causig LOC)
ODI	Open datalink interface
OH	Overhead
OOF	Out of frame
OSI	Open systems interconnection
OSIRM	OSI reference model
OSPF	Open shortest path first (Internet)
OUI	Organizationally unique identifier
P reference point	Conceptual point within a B-ISDN – the interface between the ATM switching network and a connectionless service function
P-format	A convergence sublayer format type used by AAL1
P-NNI	Private network node interface
PA	Prearbitrated segment (DQDB)
PABX	Private automatic branch exchange (nowadays synonymous with PBX)
packet	An information *block* identified by a label at layer 3 of the OSI model (e.g. X.25 packet)
packet transfer mode	A telecommunication transfer technique in which information is carried in packets
PAD	Padding
PAD	Packet assembler/disassembler
PAF	Prearbitrated function (DQDB)
PBX	Private branch exchange (an office telephone system)
PC	Priority control
PC	Printed circuit
PCI	Protocol control information
PCI	Programming communication interface (like an API)
PCM	Pulse code modulation
PCR	Peak cell rate
PDH	Plesiochronous digital hierarchy (transmission technique)
PDU	Protocol data unit
perception medium	The nature of information as perceived by the user (multimedia)
periodic frame	A transmitted element which is repeated at intervals of equal time (e.g. 125 microseconds)
PG	Peer group
PGL	Peer group leader
Ph-SAP	Physical layer SAP (DQDB)
PHP	PNNI hello packet
PHTE	PNNI horizontal topology element

PHTP	PNNI horizontal topology packet
PHY	Physical layer
PICS	Protocol implementation conformance statement
PID	Protocol identifier
PIXIT	Protocol implementation eXtra information for testing
PL	Physical layer
PL	Pad length (DQPB)
PL-OAM	Physical layer operation and maintenance cell
PL-OU	Physical layer overhead unit
PLCP	Physical layer convergence protocol (DQDB)
PLCP	Physical layer convergence protocol (defines carriage of ATM over ANSI PDH line types, e.g. IEEE 802.6)
PLK	Primary link
PLP	Packet layer protocol
PLSP	PNNI link state packet
PM	Physical medium (sublayer)
PM	Performance monitoring
PM	Performance management
PMD	Physical layer medium dependent
PMP-node	Point-to-multipoint node
PNNI	Private network–node interface or Private network–network interface
POH	Path overhead (SDH terminology)
POI	Path overhead identifier (DQDB)
PON	Passive optical network
PPP	Point-to-point protocol (Internet)
presentation medium	The means or device used to reproduce information to the user (multimedia)
PRI	Primary rate interface (ISDN)
Private local interface	Synonymous with private UNI
Private network interface	Synonymous with private UNI
Private NNI	ATM NNI interface designed for use between private ATM networks
Private UNI	ATM interface between private ATM switch (B-NT2) and ATM user (B-TE), i.e. the R or S_B (UNI) interface
PRM	Protocol reference model
PRS	Primary reference source (timing clock)
Prty	Parity
PSN	PL-OAM sequence number (PL-OAM cell)
PT	Payload type
PTE	Path terminating equipment
PTI	Payload type identifier
PTR	Pointer
PTSE	PNNI topology state element
PTSP	PNNI topology state packet
Public network interface	Synonymous with public UNI
Public UNI	ATM interface between a private ATM device or switch and a carrier network (UNI T_B or U_B)
PVC	Permanent virtual circuit

PVCC	Permanent virtual channel connection
Q3	Permanent virtual path connection
QA	TMN interface
QAF	Queued arbitrated (DQDB)
QOS	Queued arbitrated function (DQDB)
R	Quality of service
R	Reserved filed (PL-OAM cell)
R	Requirement
R reference point	Reference point at the ATM UNI between a B-TA and a B-TE2 (i.e. an X- or V-series interface)
RAI	Remote alarm indication
RBB	Residential broadband
RC	Routing control
RD	Routing domain
RD	Routing descriptor
RDI	Remote defect indication (formerly called FERF, far end receive failure)
Ref	Reference
REJ	Reject frame
representation medium	Information as described by its coded form (multimedia)
RES	Reserved
Rest 0	(Call state at the UNI) null – no transaction exists
Rest 1	(Call state at the UNI) restart request
Rest 2	(Call state at the UNI) restart
retrieval service	An *interactive* service providing for the accessing of data stored in database centres
RFC	Request for comment (IETF specification document)
RG	Regenerator
RI	Routing information
RII	Routing information indicator
RIP	Routing information protocol (Internet)
RL	Return loss
RM	Resource management
RNR	Receive not ready
RO	Read only
Root	Functional block which initiates a point-to-multipoint call
RPOA	Recognized private operating agency
RQ	Request counter (DQDB)
RS	Regenerator section (transmission)
RS	Reception status (AAL3/4 SAR-SDU)
Rs	Sustainable cell rate
RSRVD	Reserved
RT	Real time
RT	Routing infomation type
RT	Routing type
RTS	Residual time stamp
RU	Remote unit
RW	Read write
Rx	Receive
S (with suffix VCI or VPI)	Switching function for a virtual connection or virtual path

S-AIS	Section alarm indication signal (PL-OAM cell)
S-FEBE	Section far end block error (PL-OAM cell)
S-FERF	Section far end received failure (PL-OAM cell)
SA	Source MAC address
SAA	Systems applications architecture (IBM architecture)
SAA	Services and applications
SAAL	Signalling ATM adaptation layer
SAP	Service access point
SAPI	Service access point identifier
SAR	Segmentation and reassembly sublayer
SAR-PDU	Segmentation and reassemy layer protocol dataunit
S_B	Reference point at the user–network interface (UNI) between terminal equipment (B-TE) and B-NT2
SBS	Selective broadcast signalling
SBSVC	Selective broadcast signalling virtual channel
SCF	Synchronization and coordination (or control) function
SCP	Service control point (an intelligent network function)
SCR	Sustainable cell rate
SDH	Synchronous Digital Hierarchy
SDL	Specification and description language
SDT	Structured data transfer (AAL1)
SDU	Service data unit
SECB	Severely errored cell block
SECBR	Severely errored cell block ratio
SEL	Selector (address)
service attributes	Transmission capabilities and other functionality supported by an ATM or other network for the carriage of a service
service bit rate	The bitrate available to a user formation transfer
service control elements	The *primitive* controls needed to setup, manage and release a multimedia service (multimedia)
SES	Severly errored seconds
SFET	Synchronous frequency encoding technique
SIG	SMDS interest group
signalling virtual channel	A virtual channel for transporting signalling information
SIP	SMDS subscriber interface protocol (SMDS)
SIP-2	SMDS subscriber interface protocol – level 2
SIP-3	SMDS subscriber interface protocol
SIR	Sustained information rate (SMDS)
SLE	Sublayer entity
SM	Streaming mode (AAL3/4 or AALS)
SM	Security management
SM	Single mode (monomode) fibre
SMDS	Switched multimegabit data service
SMF	Single mode fibre (monomode fibre)
SN	Sequence number
SNAP	Subnetwork address plan (IEEE 802.1)
SNI	Subscriber network interface (SMDS)
SNMP	Simple network management protocol
SNP	Sequence number protection (AAL1)

SNPA	Sub-network point of attachment
SOC	Start of cell
SOH	Section overhead (SDH terminology)
SONET	Synchronous Optical Network
sound retrieval service	On-demand (user initiated) retrieval of music or sound track
Source traffic descriptor	A subset of traffic parameters belonging to ATM traffic descriptor – those requested at setup
SP	Signalling point
SP	Service provider
SP-node	Single party node
SPE	Synchronous payload envelope (SONET)
SPF	Shortest path first protocol (Internet)
SPID	Service profile identifier
SPL4	Service provider link
SPM	FDDI-to-SONET physical layer mapping standard (FDDI)
SPN	Subscriber premises network
SR	Source routing
SREJ	Select reject frame
SRF	Specifically routed frame
SRL	Structuralreturnloss
SRT	Source routing transparent
SRTS	Synchronous residual time stamp (AAL1)
SSCF	Service specific coordination function
SSCOP	Service specific connection-oriented protocol
SSCS	Service specific convergence sublayer
SSI	Service specific information
SSM	Single segment message
SSP	Service switching point (an intelligent network function)
ST	Segment type (AAL3/4 SAR)
STE	Section terminating equipment (SONET)
STE	Spanning tree explorer
STM	Synchronous transfer mode
STM	Station management
STM-n	Synchronous transport module-n
storage medium	The physical means to store information or data (multimedia)
STP	Spanning tree protocol
STP	Signalling transfer point
STP	Shielded twisted pair (cable)
STS-n	Synchronous transport signal level-n (SONET)
SUT	System under test
SVC	Switched virtual channel
SVC	Signalling virtual channel
SWG	Sub working group
SYN	Synchronous idle
synchronous transfer mode	A transfer mode which offers a periodic fixed length frame to each connection
T	Trailer
TA	Terminal adaptor

TAT	Theoretical arrival time (leaky bucket algorithm)
T_B	Reference point at the user–network interface (UNI) on the user side of the B-NT1
TB	Transparent bridging
TC	Transmission convergence sublayer (physical layer)
TCE	Transmission connection element
TCP	Transmission control protocol (Internet)
TCP/IP	Transmission control protocol/Internet protocol
TCRF	Transit connection related function
TDM	Time division multiplexing
TE	Terminal equipment
throughput	The number of bits in a block successfully transferred across a network per unit time
TIG	Topology information group
Tlr	Trailer
TLV	Type/ length/ value
TMN	Telecommunications management network
TNS	Transit network selection
TP	Twisted pair (cable)
TP	Termination point
P-AIS	Transmission path alarm indication signal (PL-OAM cell)
TP-FEBE	Transmission path far end block error (PL-OAM cell)
TP-FERF	Transmission path far end received failure (PL-OAM cell)
TPE	Transmission path endpoint
TR	Technical report
transfer mode	A telecommunications technique comprising transmission, multiplexing and switching
transit delay	The time difference between the entry of the first bit and exit of the last bit from a network
transmission medium	The physical means to transmit information (multimedia)
TRCC	Total received cell count (ATM OAM performance cell)
TS	Time stamp (ATM OAM performance management cell)
Ts	Equal to 1/Rs, the inverse of the sustainable cell rate
q	Burst tolerance
TSTP	Timestamp (ATM OAN performance cell)
TUC	Total user cell number (ATM OAM performance management cell)
Tx	Transmit
U0	(Call state at the user side of the UNI interface) null – no call exists
U1	(Call state at the user side of the UNI interface) call initiated
U10	(Call state at the user side of the UNI interface) active
U11	(Call state at the user side of the UNI interface) release request
U12	(Call state at the user side of the UNI interface) release indication
U3	(Call state at the user side of the UNI interface) outgoing call proceeding
U4	(Call state at the user side of the UNI interface) call delivered

U6	(Call state at the user side of the UNI interface) call present)
U7	(Call state at the user side of the UNI interface) call received
U8	(Call state at the user side of the UNI interface) connect request
U9	(Call state at the user side of the UNI interface) incoming call proceeding
UAS	Unavailable seconds
UB	Reference point at the UNI on the network side of B-NT1
UBR	Unspecified bitrate
UDF	User defined
UDP	User datagram protocol (Internet)
UDT	Unstructured data transferred
UME	UNI management entity
UNI	User–network interface
UNRP	User-to-network reference points
UPC	Usage parameter control
usage parameter control	The execution of appropriate action when negotiated values of information transfer are exceeded
UTOPIA	Universal test and operations physical interface for ATM (ATM specification for standard ATM/PHY layer interface)
UTP	Unshielded twisted pair (cable)
UTP3	Unshielded twisted pair cable, category 3
UTPS	Unshielded twisted pair cable, category 5
UUS	User-to-user signalling
valid cell	A cell in which the header is declared by the header error control process to be free of errors
variable bitrate service	A telecommunication service using a service bit rate based upon statistical parameters
VBR	Variable bit rate
VC	Virtualchannel
VC CEPF	virtual connection connecting end point functions
VC CPF	Virtual connection connecting point functions
VC-n	Virtual container-n (SDH terminology)
VCC	Virtual channel connection
VCCE	Virtual channel connection endpoint
VCI	Virtual channel identifier
VCL	Virtual channel link
VF	Variance factor
video on-demand	A video *retrieval* service
videomessaging	A *messaging* service for transfer of video material
virtual channel connection	A concatenation of virtual channel links
virtual channel link	ATM connection between a point where a virtual channel identifier is assigned and a point where it is removed
virtual path connection	A concatenation of virtual path links
virtual path link	ATM connection between a point where a virtual path identifier is assigned and a point where it is removed
VoD	Video on demand

VP	Virtual path
VP CEPF	Virtual path connection end point functions
VP CPF	Virtual path connecting point functions
VP-XC	Virtual path crossconnect
VPC	Virtual path connection
VPCI	Virtual path connection endpoint
VPI	Virtual path identifier
VPL	Virtual path link
VPRPC	Broadband virtual path service for reserved and permanent connections
VPT	Virtual path terminator
VT	Virtual tributary (SONET)
VTOA	Voice over ATM
WAN	Wide area network
WTSC	World Telecommunicaton Standardization Conference
X	Interface between two managment systems (TMN)

Index